THE MOUNTAIN WEST

Creating the
North American
Landscape

Gregory Conniff

Bonnie Loyd

Edward K. Muller

David Schuyler

Consulting Editors

Published
in cooperation
with the Center
for American
Places,
Harrisonburg,
Virginia

THE MOUNTAIN WEST

Interpreting the Folk Landscape

Terry G. Jordan

Jon T. Kilpinen

Charles F. Gritzner

The Johns Hopkins University Press

BALTIMORE & LONDON

© 1997 The Johns Hopkins University Press
All rights reserved. Published 1997
Printed in the United States of America
on acid-free paper

06 05 04 03 02 01 00 99 98 97   5 4 3 2 1

The Johns Hopkins University Press
2715 North Charles Street
Baltimore, Maryland 21218-4319
The Johns Hopkins Press, Ltd., London

Library of Congress Cataloging-in-Publication Data
will be found at the end of this book.
A catalog record for this book is available from the
British Library.

All maps prepared by John V. Cotter

Title page illustration: Anglo-Western cabin in Alaska with cantilevered porch, 1899. Courtesy of the Alaska State Library, Juneau, Winter and Pond Collection, no. PCA 87-1270.

ISBN 0-8018-5431-8

TO FRED KNIFFEN, 1900–1993

Founding father of research in American folk geography
and personal inspiration for all three of us. We think he would
have liked this book and regret only that it came a few years
too late. His kind of geography is the best of all.

The source of all geographical knowledge is in the field.

COTTON MATHER

Contents

Acknowledgments

In addition to our intellectual and methodological debts to Fred Kniffen, we ran up a whole lot of personal obligations during our three decades of roaming through the West in search of "neat old stuff." With three or four really notable exceptions, the western people we encountered proved to be salt of the earth. Many among them provided substantial help in our field research, directing us to noteworthy structures and useful collections while sharing their insights and intimate knowledge of different places or regions.

These include Dan Turbeville, of Eastern Washington University; Jim Rock, of Klamath National Forest; Katie Thorsheim, of Grand Lake and Golden, Colorado; Jim Scott, of Western Washington University; Mr. and Mrs. E. Parker, the curators at 108-Mile Heritage Site in British Columbia; Norman Staton, owner of the Log Cache in Sitka; Linda Yarborough, of Chugach National Forest; and Steve Jett, of the University of California, Davis. Dan Turbeville and Jim Scott each on two different occasions took time from their work schedules to accompany us into the rural areas of the Pacific Northwest, and Katie Thorsheim was similarly generous. Numerous ranchers allowed us onto their property—no small gift in this age of crime and suspicion. Only once were we sent on our way with a dog at our heels. Even long-suffering American Indians, who have every reason to avoid us, proved hospitable.

We also salute those relatively few westerners who, perceiving the endangered condition of the region's old wooden structures, have undertaken historical preservation in the form of small outdoor museums. They labor in a lonely vineyard, trying their best to keep alive shards of the traditional western land-scape, often in trying circumstances and usually without much financial help. Many of these museums are listed among the sources at the end of the book.

George F. Thompson, president of the Center for American Places in Harrisonburg, Virginia, has encouraged and facilitated this project from its inception, and we are most indebted to him. Dealing with three troublesome authors on a single book must be trying, but he has never lost his cheerful disposition.

John V. Cotter, of Austin, Texas, provided the excellent cartography for the book. We had initially sought his services for the project, only to find him too swamped with work to take it on. Another, less talented map-maker did maps that proved unsatisfactory. We begged John to redraw everything in his own special, old-fashioned way, and we are very grateful that he found the time to do so. The book would be far more ordinary had he not placed his mark on it.

THE MOUNTAIN WEST

Chapter One # THE NORTH AMERICAN WEST *Continuity or Innovation?*

The most fundamental cultural division of North America, many would agree, has always been between East and West. Everyone's mental map has a place for the West, with associated images. An elaborate mythology and iconography have become associated in the popular mind with that huge section of the continent, and "the American West lives on in the national imagination," a reservoir of "collective fantasies." Myth, image, and fantasy find their basis in a reality: the West is not like the East in a variety of measurable ways, in climate, terrain, religion, politics, and the propensity to purchase Elvis Presley memorabilia. A distinctive subculture resides out there.[1]

Nor is this sectionalism confined to the United States. Canada, too, has a West, both in fact and perception. In many ways the Canadian and American Wests have more in common with each other than either does with its respective East. Beyond the 100th meridian, international borders become more difficult to perceive. Our study will, accordingly, ignore the borders, ranging from the fringes of Latin America to the wilds of the Yukon and Alaska.

Using an unorthodox methodology applied for the first time to the West at large, we seek in this study to assess the degree of western cultural distinctiveness and, more important, to determine the mainspring cause of this fabled sectionalism. We believe we have found material evidence, a key if you will, to unlock the mysteries of the West. That evidence resides in the thousands of traditional wooden buildings still found on the western landscape.

Explaining the West

Scholars from a variety of academic disciplines have repeatedly
addressed the very questions we now pose, without achieving
any consensus. They have put forward two fundamentally different and contradictory explanations for western sectionalism: the
West as the displaced, archaic, frontier East and the West as a
subculture developed indigenously to meet the challenge of unique and difficult environmental conditions.

The most persuasive case for the West as indigenous was offered by Walter Prescott Webb, whose formative rural years
were spent virtually astride the 100th meridian, that symbolic
divide between West and East. He proposed that the physical environment of the West lay at the root of it cultural character. An
eastern culture, shaped in a humid, forested land, simply could
not cross unaltered into the West, much of which was moisture-
deficient and treeless. Fundamental changes had to be made
when pioneers crossed the 100th meridian—changes in institutions and means of livelihood. Innovations such as barbed wire
and the water-pump windmill were prerequisite to life on the
land. Laws had to be rewritten, adaptive strategies altered,
architecture redesigned, and transport systems restructured.[2]
Webb's work proved enormously influential, among historians at
least, with the result that "most of those who write western history seem to assume that physical environment has dominated
western life and has made the West rough and radical," in contrast to a genteel and conservative East.[3] Webb and his imita-
tors, then, argued that the West owed its cultural distinctiveness
to *innovation* and represented a sectional culture *born in place*,
largely as a result of a massive cultural readaptation to a new
and difficult physical environment. Many or even most lay westerners share this view. That may help explain the continuing
popularity of Webb's book, *The Great Plains*, which still sells well
in the West. Talk to westerners—the *real* ones out there on the
land, not the plastic purveyors of popular culture inhabiting
cities such as Los Angeles or Calgary—and you will often hear
them say something to the effect that the West is too physically
harsh and demanding for easterners. In any case, most westerners and many students of the West have long regarded the regional culture as indigenous.[4]

The contrasting explanation holds that the West differs mainly by virtue of better preserving a pioneer culture now largely
vanished from the East. The West is the archaic East. Alaska
bills itself for touristic purposes as "the last frontier," not entirely
without justification. Ralph Waldo Emerson called Mormonism
"an afterclap of Puritanism," and the Mormon culture region in
the Great Basin did, indeed, resemble a latter-day colonial New

England, complete with theocratic puritanism and farm villages.[5] Frederick Jackson Turner, contrary to the popular image of his works, also wrote compellingly at times in support of the displaced East thesis. "Stand at Cumberland Gap," he advised, "and watch the procession of civilization, the fur trader and hunter, the cattle-raiser, the pioneer farmer. Stand at South Pass in the Rockies a century later, and see the same procession."[6] Echoing Turner's sentiments, Frederick Simpich spoke of "Missouri, mother of the West," while Earl Pomeroy chastised mainstream historians for "neglecting the spread and continuity of Eastern institutions and ideas," proposing instead that the westerner was an "imitator rather than innovator" who relied upon an "inheritance" from the East.[7]

The so-called new historians of the West, we feel, have added little to the fundamental debate concerning the cultural roots of the region. Both Turner and Webb have taken a beating by the new historians, but to what explanatory effect? Anti-Turnerian rhetoric, long an intellectual cottage industry for the discipline of history, serves merely to drive some of the new historians toward a Webbian neo-environmental determinism. Donald Worster, for example, views the West much as Webb did, as a region delimited by aridity. Others replace Turner's "frontier" with "conquest"—environmental, cultural, political, and economic—as the defining event that produced the American West, but conquest also occurred in the East and we wonder how the same process in the two halves of North America could have yielded such sectionalized results.[8]

Still other historians, laboring in a second venerable cottage industry, continue to defend Turner or to lend him unintentional support by documenting cultural continuities between East and West. But most simply avoid the sticky question of western cultural origins and are content to tell us, correctly enough, that Indians were mistreated, the land raped, and the role of women underestimated in the settlement of the West.[9]

Every now and then, some fresh air blows over the subject of explaining western culture. Folklorist Jennifer Attebery, for example, sought to join the Turnerian and Webbian models, proposing that certain aspects of western culture were "patterned after eastern models but adapted to western conditions."[10] Regrettably, no such speculations have so far led to definitive statements concerning western sectionalism.

Geographers, bringing the spatial perspective of their discipline to bear on the question, have suggested that the West is not monolithic, but instead "an aggregate of distinctive subregions." To comprehend it, we need to understand its component parts. The mountain West, largely wooded and well-watered, is one such subregion.[11] To this spatial perspective, we now add a traditional geographic methodology, in an effort to understand the West: cultural landscape analysis.

Cultural Landscape

For generations, geographers laboring in the humanistic tradition of their diverse field have believed that the archaic greater artifacts found in ordinary, human-built landscapes—folk dwellings, old barns, relict fences, weathered granaries, and the like—possess a diagnostic power to chart diffusion, reveal innovation, display adaptive systems, and explain regional cultures (Fig. 1.1). Old buildings, carefully and correctly analyzed, bear worthwhile messages that can provide insight and even revelation.

This venerable cultural landscape methodology runs deep in geography, through Fred Kniffen and Carl Sauer and ultimately back to the originator of landscape analysis, the nineteenth-century German scholar August Meitzen, who told us, truthfully, that we walk "in every village among the ruins of antiquity."[12] Later generations of cultural geographers agreed, and Peirce Lewis proposed that, if properly sensitized, we could "read the landscape as we do a book."[13] Edwin Brumbaugh went so far as to say that folk architecture encompassed "all that is really worth telling about a people" (Fig. 1.2).[14]

Neither "reading" nor "telling" come easily, however. Cultural landscapes reveal little meaning unless each component is observed and understood both singly and in the context of its functional integration into the local settlement complex and regional adaptive strategy of land use (Fig. 1.3).[15] In short, if settlers entering the West found that strategies and techniques introduced from their former homelands succeeded in the West, then their built environments—their artifacts—will reveal that success. If the introduced adaptive strategies did not work in the West, then the resultant innovations and changes are also abundantly imprinted in the cultural landscape. The geographer, by training, learns to place individual artifacts and humanized landscape complexes into adaptive context, achieving in the process an explanatory power.

This book represents, to our knowledge, the first attempt to bring this revelatory aspect of the cultural landscape to bear upon the problem of understanding the West. We use the greater artifacts of western material culture to address the venerable issue of western cultural distinctiveness and regionalism, to resolve the basic question of indigenous innovation versus diffusion from outside.[16] We bring our geographical perspective and method to the problem of western identity.

Figure 1.1. Anglo-Western, or "Finnish plan" single-pen log cabin, probably built by a trapper or prospector, west of Dawson City, Yukon Territory (field research district 21). The 14-by-16-foot front-gable dwelling has a turf-over-bark roof supported by a ridgepole-and-purlin structure. (Photo by T.G.J., 1990.)

Figure 1.2. Truchas, in Hispanic highland New Mexico, possesses a remarkable array of notched log buildings (research district 2). (Photo by T. Harmon Parkhurst, courtesy of the Museum of New Mexico, Santa Fe, neg. no. 11592.)

Figure 1.3. A complex of three log outbuildings on a wheat farm in Meagher County, Montana (research district 8). The West offers countless such traditional clusters of log buildings, few of which have ever been studied. The nearest structure is a single-crib barn. Two of the roofs are raftered and one is a ridgepole-and-purlin type. (Photo by T.G.J., 1990.)

Methodology

Implicit in the cultural landscape methodology is a heavy emphasis upon field research. Emphatically rejecting historian Earl Pomeroy's claim that "the sources on the West are in large part literary," we have, by and large, slighted libraries and archives.[17] Instead, we present here observations and conclusions based upon literally hundreds of first-hand inspections of relict western buildings. Dirty boots and wet socks go with cultural geography, and we know more about sunburn and sore feet than we do of hemorrhoids. We distrust data gathered in any other manner, placing us in closer kinship to anthropologists, folklorists, photographers, and artists than to historians. We are inclined to disdain social scientific theory, an opiate which only clouds the mind and distorts the observation. You will find our work rather thinly footnoted for academic writing, since relatively few students of the West have left the library and preceded us in field research.

We three ranged across the West off and on during the past thirty summers, happily seeking out "neat old stuff," measuring, taking photographs, inspecting details of construction, conversing with the people who use the structures, enjoying the aesthetic pleasures of open spaces, and indulging in unbridled bucolic romanticism. There are worse ways to spend one's best years. Our research took us from New Mexico's highlands to Alaska's Arctic Circle, a span of country stretching nearly 3,000 miles.

We imposed no particular border upon the West, knowing as geographers that such cultural boundaries are never sharp and that drawing such limits as lines on maps represents a fool's errand. We just drove west until we knew we were there.

The sheer size of the region under consideration, regardless of how you chose to bound it—half a continent—required sampling, achieved in several ways. We chose to focus upon traditional wooden buildings, in particular those involving log construction (Fig. 1.3). Wood lacks permanence, and the endangered character of these structures gave us an added sense of urgency.[18] If we were to learn something about the West from log cabins, wooden barns, and pole fences, then the hour was late. Obeying our instincts as geographers, we further restricted our attention to the mountain West, choosing this single distinctive subregion for analysis. Then we selected twenty-five smaller districts within the montane area for detailed field research (Fig. 1.4). We achieved in this way a broad-based geographical sample of the mountain West. Field observations made outside the twenty-five districts appear in our book, but by and large we concentrated our labors in those places. In all, we inspected some 1,550 log structures in the twenty-five districts and about 500 additional examples in other, intervening areas.

We did not, then, cover the entire West geographically or culturally, but instead sought to learn something meaningful about selected aspects of particular parts of the region utilizing one method, hoping that what we observed represented some substantial part of the whole and that our conclusions possessed a wider applicability. We do not apologize for approaches not taken, for methods spurned, for places unvisited, or for topics ignored. We ask that you accept our work for what it is—a straightforward reading and interpretation of a fragment of a regional folk landscape, devoid of myth, unstructured by theory, and untainted by science. We did not attempt to create a definitive work on the West, though anyone seeking such an Olympian view would do well to listen to what we learned from some old wooden buildings.

We have, in short, accomplished the first objective, demythologized analysis of a regional culture in the West based upon a tangible, standardized, and statistically measurable data base. People and Professor Pomeroy's literary sources spin tall tales and myths as often as not; the greater artifacts, sensitively and accurately interpreted, invariably tell the truth. They are the substance. Listen, now, to the messages of the cultural landscape.

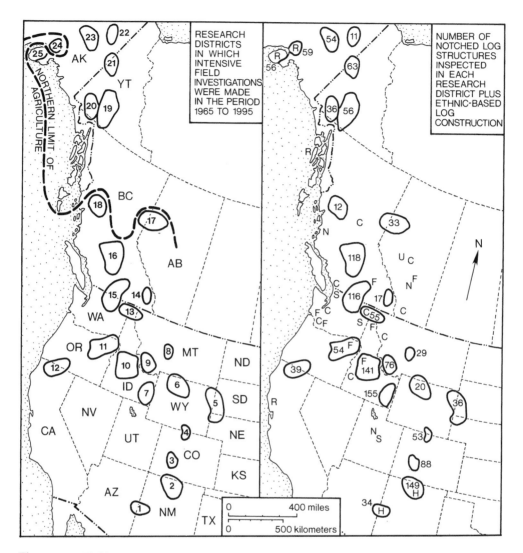

Figure 1.4. Field research districts, ethnic enclaves, and the number of log structures inspected in each. The research districts are: 1. Mogollon Rim; 2. Highland New Mexico; 3. Gunnison River area; 4. North Park–Upper Platte; 5. Black Hills–Pine Ridge; 6. Big Horn Basin; 7. Mormon Idaho; 8. Little Belt Mountains; 9. Beaverhead Country; 10. Salmon River Country; 11. Blue Mountains; 12. Southern Cascades; 13. Kaniksu–Selkirk Mountains; 14. East Kootenay Trench; 15. Okanogan Country; 16. Cariboo-Chilcotin District; 17. Peace River Country; 18. Skeena–Coastal Mountains; 19. Upper Yukon River Valley; 20. Kluane Country; 21. Middle Yukon River Valley; 22. Arctic Circle–Steese Highway; 23. Middle Tanana River Valley; 24. Mat-Su Valley; 25. Kenai Peninsula. The ethnic groups are: C = Canadian French; F = Finns; H = Hispanos; N = Norwegians; R = Russians; S = Swedes; U = Ukrainians. These ethnic locations refer only to group settlements. Isolated individuals living outside the colonies shown also engaged in log carpentry, at diverse locations not shown.

Chapter Two   # LOG DWELLINGS

August Meitzen, the founding father of cultural landscape study, exalted the folk house as "the embodiment of a people's soul."[1] One might well argue that people erect no structures more revealing than their homes. The traditional log dwellings of the montane West, then, ought to tell us a lot about the regional culture and character.

Three generations of research on the folk houses of the eastern United States and Canada have yielded hundreds of books and articles on the subject. As a result, we know in some considerable detail about the multiple regional housing traditions found back East. These include the *Tidewater Southern* type, dominant along the coast between Chesapeake and Galveston Bays; the *French Canadian* tradition, centered in riverine Québec; the *Yankee New England* type, which also laps over into the Canadian Maritime Provinces and Ontario; and *Midland* or Pennsylvania folk housing, dominant in a huge wedge-shaped region with an eastern apex in the lower Delaware River Valley and reaching west through the Upland South and lower Midwest to the margins of the Great Plains.[2]

Oddly, comparatively little has been written about the folk houses of the West. Perhaps the general assumption has been that Anglo-Americans carried their housing traditions with them along the trails west. In this chapter we present a comparison of eastern and western dwellings, a comparison that will go a long way toward resolving the issue of diffusion versus innovation in the West. At the outset we can say that the log folk house types most common in the American and Canadian East occur relatively infrequently in the montane West (Fig. 2.1).

Figure 2.1. A rare mountain western dogtrot cabin, in the sagebrush expanses of North Park, Jackson County, Colorado (research district 4). Built of V-notched round logs hauled from the nearby mountains, the derelict cabin has a ridgepole-and-purlin roof covered with boards. Here, as in the East, the dogtrot generally involved the joining of two Anglo-Western cabins. (Photo by T.G.J., 1987.)

Double-Pen Log Houses of Eastern Origin

Three major types of *double-pen*, or two-room, log houses prevail in the East, each linked to the Midland tradition. If the paired rooms flank an open central breezeway, the result is the well-known *dogtrot* house. Another favorite arrangement places the pens on either side of a central chimney with double fireplaces, a type known as the *saddlebag* house. In the third Midland type, the pens abut one another, without either a central passage or chimney, a variant called the *Cumberland* house. All three of these eastern double-pens form elongated dwellings, with a broad eave side facing forward and often multiple front doors. Each could be enlarged to full two-story height, creating the so-called *I-house*.[3]

None of these eastern double-pen houses achieved a noteworthy presence in the West. We looked at about 850 log dwellings in the twenty-five field research areas, finding only fourteen dogtrots, eight Cumberlands, one saddlebag, and no I-houses (Fig. 2.2). Beyond our specific areas of study, we happened upon nine additional dogtrots, four Cumberlands, a few saddlebags, and one I-house. In our sample, then, eastern double-pen types accounted for less than 3 percent of all log dwellings.[4] Reputedly, a Cumberlandlike double-pen plan, without fireplaces or chimneys is "particularly common in the Rocky Mountains," but we found evidence to the contrary (Fig. 2.3).[5]

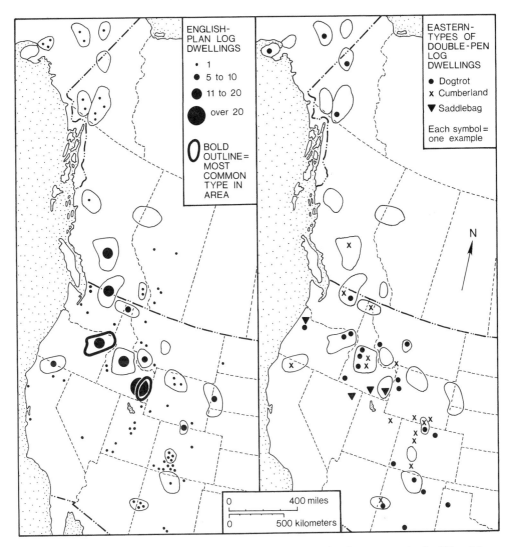

Figure 2.2. Distribution of eastern log house types in the West. Most do not have chimneys or fireplaces, and the majority of English-plan cabins lack a rear door and are more elongated than those found in the East.

Several factors may help explain the failure of the eastern types to become common in the West. The open-air breezeway of the dogtrot turned into a frigid wind tunnel in the mountain western winter, and the eastern chimney-fireplace gave way to iron stoves before most western settlement began, rendering the saddlebag obsolete. The eastern double-pens served an agrarian, family-based population, whereas the most common western housing sheltered unmarried males. Moreover, the eave-front arrangement of the eastern double-pens presented a problem in a snow-rich climate, creating a drip-line quagmire at the en-

trance during the thaw season and encouraging snowdrifts to ac-
cumulate in winter.

That same problem associated with eave-front houses perhaps
helps explain the virtual absence of most Yankee, Tidewater, and
French-Canadian dwelling types in the West. Indeed, the discov-
ery of a lone New England–style log house in California was suf-
ficiently exciting to prompt a scholarly article.[6]

English-Plan Cabins

The most common eastern log house, perhaps accounting for
more than half of those ever built east of the Mississippi, is a
single-pen, or one-room structure with side-facing gables and a
door in each eave wall (Fig. 2.4). It occurs in three of the four
folk housing traditions of the East, excluding only the French-
Canadian. Sometimes referred to as the *English plan* house, this
small dwelling is roughly square in plan, with an average dimen-
sion of about 18 feet on a side. In some cases, the eave walls may
be up to 4 or 5 feet longer than the gable sides. A more elongated
but less common subtype, sometimes called the *hall-and-parlor*
or *Scotch-Irish*-plan cabin, has eave walls about 8 to 10 feet longer than those on the gable and is often subdivided internally to
form separate rooms of unequal size.[7] These eastern eave-entry
log houses have a chimney and fireplace centrally positioned in
one gable wall, providing both heat and an open hearth for cooking.

Figure 2.3. Western double-
pen similar to the Cumberland
type of the East, lacking only
the end fireplaces and chimneys. This type, here seen in
Okanogan County, Washing-
ton (research district 15), is
very rare in the West. (Photo
by J.T.K., 1990.)

Figure 2.4. A story-and-a-half English-plan (or eave entrance) cabin, built in 1909 and serving as the Slate Creek Ranger Station, Idaho County, Idaho (research district 10). The roof is raftered and the corners are square notched. Unlike most western English-plan cabins, this one has a rear door. (Photo by T.G.J., 1987.)

The English-plan cabin succeeded better than the eastern double-pen plans in gaining a foothold in the West (see Fig. 2.22). We found 139 such houses in our twenty-five districts, amounting to about 16 percent of all western log dwellings, and they appeared consistently in all but some of the northernmost research areas (Fig. 2.2). In only a few districts, most notably Mormon Idaho, did English-plan houses constitute the dominant type. Among the Mormons, almost three of every five log dwellings were of this type. Perhaps the Mormons, whose cultural roots reached back to English-dominated colonial New England, retained an ancestral preference for the plan. Once established in the Great Basin, the Mormons recruited many new converts in England, bringing them to Deseret and reinforcing English tastes. Similarly, the western Canadian population derives largely from English ancestors, possibly explaining the above-average frequency of English-plan cabins in the Cariboo-

Chilcotin (19 percent), East Kootenay Trench (25 percent), and Okanogan country (21 percent) of British Columbia and adjacent Washington.[8]

The distinctive Mormon tradition regarding house plan alerts us at the outset of our study to a truth geographers know instinctively, namely, that the West, even the Anglo-American West, is not culturally monolithic. The landscape has warned us, in the varying distribution and level of acceptance of the English-plan log cabin, that explaining the West cannot be achieved singly and simplistically. The region constitutes no organic whole with a single sectional character. Rather, an array of unique western regions await explanation.

Returning to our humble English-plan cabin, we find that the majority of western examples exhibit an elongation, somewhat in the Scotch-Irish manner, in contrast to frequencies observed in the East. But the partition of this elongated plan into separate hall and parlor, found in the eastern Scotch-Irish type, is usually absent in the West. In addition, most western examples lack a chimney-fireplace, and the rear door is usually also absent.

The greater question concerns the large-scale abandonment of the English-plan house by most western Anglo-Americans. Its chief disadvantage, in all probability, lay in the eave entrance, which suffered the same snowdrift and meltwater problems afflicting the eastern double-pen plans. The absence of a rear door in the typical western English-plan cabin probably reflects these same problems.

The Anglo-Western Cabin

Taking the place of the most common eastern log houses as the dominant type in the West is a small, front-gable, single-pen cabin (Fig. 2.5; see also Fig. 1.1). Variously called the "Anglo-Western," "Rocky Mountain," and "Finnish-plan" house, it accounts for about 54 percent of the dwellings we recorded. In all but three of the research areas (Mormon Idaho, highland New Mexico, and Oregon's Blue Mountains) the Anglo-Western cabin proved to be the most common type (Fig. 2.6). In a few districts, such as the Beaverhead country of Montana or the middle Tanana Valley in Alaska, Anglo-Western cabins exceeded two-thirds of the total. Other studies confirm similar dominance in certain additional western districts.[9]

The Anglo-Western cabin has a single door positioned in the front gabled wall, usually with one small window in each side wall. Normally rectangular in shape, these dwellings measure 10 to 15 feet wide at the gable ends by 10 to 22 feet deep along the axis of the roof ridge. Gable-end entry, the key diagnostic feature, allowed low eaves, permitted hillside dugout siting if that was desired, and facilitated addition of a cantilevered roof pro-

Figure 2.5. An Anglo-Western log cabin, the most common sectional type, in Alaska. Note the round-log construction and ridgepole-and-purlin roof covered with shingles. The side eaves of these front-gable cabins typically hang low. (Photo courtesy of the Alaska State Library, Juneau, Winter and Pond Collection, no. PCA 87-2567.)

jection to shelter the entrance (Fig. 2.7). The cantilevered porch is very common throughout the mountain West, although application of roofing to cover the projection was, curiously, sometimes delayed for years, presenting the odd sight of nonfunctional projecting beams (Fig. 2.8).[10]

The Anglo-Western cabin served as the most common dwelling for prospectors, trappers, cowhands, lumberjacks, forest rangers, and homesteaders alike throughout most of the West (Fig. 2.9). It also became the dominant type of Amerindian and mixed-blood housing from Wyoming and Idaho northward to Alaska and the Yukon. The native American quarters of many towns and villages in the far North consist of a row of Anglo-Western log cabins lining the banks of rivers. So abundant and dominant is this house type that it serves as a regional symbol and icon, a way of saying "we are westerners." Any number of posters, postcards, advertisements, and wall murals in Alaska

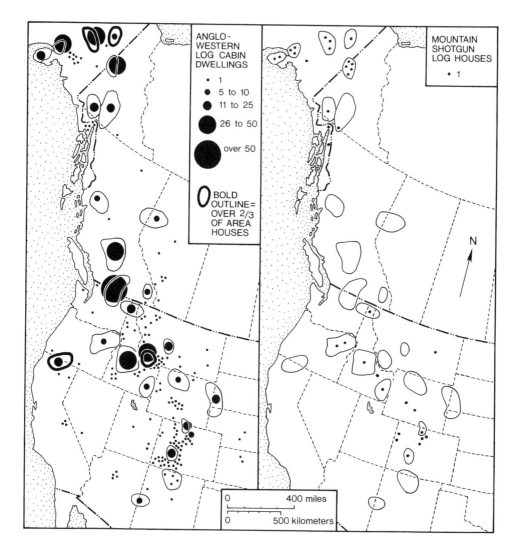

Figure 2.6. Distribution of two front-gable cabin types in the West. The Anglo-Western cabin represents by far the most common western type, while the mountain shotgun cabin outnumbers all other double-pen plans.

and the Yukon depict the gable-entry log cabin side-by-side with images of dogsleds and mountain men.[11]

Clearly, then, the Anglo-Western cabin represents a type closely linked to the region, a symbol of sectional culture and distinctiveness. We might easily accept it as a western innovation or adaptation, readily explained by climatic advantages. Not only does it accommodate snowfall and thaw with minimal difficulties, but also, by virtue of entry through a gable end wall, which stands taller, the structure can be lower to the ground. For this

Figure 2.7. An Anglo-Western cabin in Alaska with cantilevered porch, a common variant of the front-gable plan. The porch served to shield the entrance from snowdrifts and meltwater. The round logs, with bark intact, are V-notched. Smoke pours from a corner stovepipe. A shingled ridgepole-and-purlin roof covers the cabin. (Photo 1899, courtesy of the Alaska State Library, Juneau, Winter and Pond Collection, no. PCA 87-1270.)

Figure 2.8. Ridgepole and purlins project from this Alaskan Anglo-Western cabin, in anticipation of a cantilevered porch. Such arrested construction, oddly, is not uncommon. (Photo courtesy of the Alaska State Library, Juneau, Harry T. Becker Collection, no. PCA 67-126.)

Figure 2.9. A very humble Anglo-Western cabin, home to an obviously discouraged California prospector. Note the ridgepole-and-purlin
roof and crude chinking. (Photo courtesy of the California State Library, Sacramento, Photograph Collection, neg. no. 20,609.)

reason the Anglo-Western cabin requires fewer rounds of logs to
build and enjoys lower-hanging protective eaves.[12]

Closer analysis, however, reveals the logic of western origin to
be false. Eastern states offer abundant examples of the gable-
front floorplan in the Midland folk architectural region, whether
as dwelling, barn, kitchen, or smokehouse. Indeed, the front-
gable plan existed in the East as far back as the 1650s, when Finnish and Swedish settlers colonized the lower Delaware River
Valley.[13] Front-gable log cabins served as troop quarters in both
the Revolutionary and Civil wars, and houses of this type can still
be found occasionally in the East. Additional support for an eastern origin of the front-gable cabin is provided by its remarkably
widespread distribution in the West. Innovations developed in
the West, as we will demonstrate later, rarely if ever attained a
pan-western distribution, but instead remained concentrated in
particular parts of the West and occurred most commonly near
their place of invention.

In the Midland East, the front-gable cabin plan never ap-
proached the dominance it later enjoyed in the West. How could
a small minority type in the parent East come to prevail in the
western child? Cultural ecologists instruct us that adaptive strategies of colonization and land use typically include diverse rarely

used techniques and material culture forms. In other words, some elements of Anglo-American folk architecture in the eastern states appeared only here and there on the cultural landscape. They formed part of the collective folk memory and building repertoire, but were rarely employed. The front-gable cabin was such an element. Research often reveals these unfavored forms to be archaic. When a people enter a new physical environment, such as the West, the archaic forms sometimes prove better preadapted for occupance of the land than the favorite types, leading to a renaissance. This ecological process, we feel, operated in the case of the front-gable cabin, which blossomed into regional dominance in the West, where it enjoyed ecological advantages.

In addition, the front-gable plan still today characterizes the most common type of Midland eastern log barn, a simple granary with hayloft above. The plans of barns and other outbuildings often perpetuate archaic house types, suggesting that the eastern gable-entrance single-crib log barn may represent a dwelling form that was more common in pioneer days and became archaic as the frontier passed. In some measure, the front-gable cabin of the West represents the archaic East.

The message of this modest log house type concerning the cultural roots of the West, then, is anything but simple. It tells us the archaic East lives on more abundantly in the West and also that certain eastern forms proved better preadapted than others to the western climate. All eastern log house types that diffused as far westward as the Mississippi Valley subsequently appeared in the montane West, but only a few thrived there. A stern selection process operated, causing the basic order among single-pen cabins to shift from English to Anglo-Western plan.

Mountain Shotgun Cabins

The failure of all three eastern double-pen plans to succeed in the western states prompted an innovation. A distinctively western enlargement of the Anglo-Western plan produced the *mountain shotgun* cabin, in which a second log room abuts the rear wall of the original pen (Fig. 2.10).[14] Because of the difficulty of splicing log pens together, mountain shotguns often have "telescoped" walls, with the rear pen narrower than the front room, its walls tucked inside (Fig. 2.11). Only if the two rooms were built at the same time, employing a single span of logs, could this problem be avoided. The mountain shotgun can easily be further enlarged, to three or even four rooms (Fig. 2.12).

While the mountain shotgun cabin is the most common double-pen type in the West, relatively little demand apparently existed for such large dwellings. As a result, the type is not especially common. We found only twenty-one examples in the re-

Figure 2.10. Mountain shotgun double-pen cabin, on Ohio Creek, Gunnison County, Colorado (research district 3). This type probably evolved in the West from the Anglo-Western cabin. (Photo by J.T.K., 1989.)

Figure 2.11. Telescoped side walls on shotgun cabin, built about 1910 at McConnell Ranch, Robinson, Yukon Territory, between Whitehorse and Carcross (research district 19). Note the pole and turf roof, and the weight poles. The cabin measures 15 by 38 feet. As a result of telescoping, the rear pen consists of only three walls and measures 14 feet wide at the rear. (Photo by T.G.J., 1990.)

Figure 2.12. A mountain shotgun four pens deep at Eagle on the Yukon River, Alaska (research district 21). Built about 1900, it well reflects the shotgun's expansion potential. (Photo by T.G.J., 1990.)

search areas, about 2.5 percent of all log houses, and we happened upon six others. Well over half occur in Alaska and the Yukon Territory, beyond the northern limits of agriculture (Fig. 2.6). In the middle Yukon Valley region, shotguns accounted for one in every ten houses. Equipped with a false front, the shotgun plan often serves as a store, as can be seen at Bannack, Montana, and many other places. We did not count these as dwellings.

The term "mountain shotgun" acknowledges that a house type of virtually identical plan and name, though not of log construction, appears widely in the Tidewater South, especially among persons of African ancestry. This southern shotgun shack seems to us an unlikely prototype for its mountain western counterpart. Innovative, indigenous development from the front-gable single-pen cabin appears more likely. For those relatively few western settlers who desired a double-pen house, the mountain shotgun offered all of the ecological advantages of the parent single-pen, while avoiding all the parallel disadvantages of eastern double-pens.

Were it not for the thin scattering of mountain shotguns south of the Canadian border, we could comfortably regard Alaska or the Yukon as the place of origin of the type. Perhaps multiple invention occurred, and we cannot wholly dismiss the possibility of a link to the Tidewater shotgun. All we can say conclusively is that the most common double-pen dwelling in the West did not occur in the log building tradition of the East.

Figure 2.13. Enlarged, story-and-a-half front-gable single-pen log house, Talkeetna, Alaska (research district 24). Modeled upon the much smaller Finnish-plan cabin, these dwellings are so much larger that they constitute a different, distinctively western type. Also, internal subdivision into rooms has occurred. End-only hewing and square notching can be seen. (Photo by J.T.K., 1990.)

Other Distinctively Western Log Houses

An even better case for western innovation is provided by the 230 dwellings, roughly a quarter of the total, that match none of the types mentioned so far. In each case, these represent substantial or total departures from eastern tradition.

One favored western practice, characteristic of the postfrontier era, was to increase greatly one or, more typically, both dimensions of English-plan or Anglo-Western cabins, producing versions so oversized as to require a new typology (Figs. 2.13 and 2.14). Such enlargements, made possible by the tall, slender timber available in much of the West, added 5, 10, or even more feet to the exterior dimensions, and often an upper story as well. In our research areas, we found forty-seven oversized front-gable houses and twenty with eave entry, together making up about one-fifth of the nonconformist dwellings. Like the mountain shotgun, the enlarged gable-entry type proved well suited as a false-front store (Fig. 2.15).[15]

Many additional Anglo-American western log houses simply defy classification. They exhibit diverse, unique, rambling floorplans that follow no eastern model but also fail to be duplicated locally in the West (Fig. 2.16). A house "type" demands multiple examples, a criterion not met here. Instead, almost every imaginable random configuration appears here or there. One New Mex-

Figure 2.14. Enlarged eave-front single-pen house, near Del Norte, Rio Grande County, Colorado. While similar to the English-plan house of the East, this dwelling is much larger, with internal subdivision by board walls into separate rooms. The result is a distinctively western plan. (Photo by T.G.J., 1987.)

Figure 2.15. False-front store in Challis, Custer County, Idaho (research district 10), consisting of an enlarged front-gable log structure. The building, in fact, is an oversized, two-story mountain shotgun. (Photo by T.G.J., 1987.)

Figure 2.16. An odd, unclassifiable log house near Lander, Natrona County, Wyoming. Such rambling, nonconformist dwellings, each displaying unique characteristics, abound in the West, implying individualism and the weakness of folk culture. (Photo by T.G.J., 1987.)

ico log ranch home, for example, had twenty-six rooms![16] Such houses perhaps best reflect both innovation and nonconformity in the West. Some 163 dwellings fall into this category, 19 percent of the total. The districts showing the highest proportions of these unclassifiable log houses are highland New Mexico, with 81 percent, and the Peace River country, with 50 percent. Indeed, the New Mexican highlands form the most distinctive and complex culture area in our study.

Ethnic Influences

Ethnicity rather than western individualism lies at the root of these highest levels of deviation in dwelling design. In highland New Mexico, the great majority of houses inspected were built by Hispanos, an agrarian people rooted in this land since the 1600s and perpetuating a log building tradition derived from Mexico rather than the eastern United States. The traditional Hispanic log house in New Mexico was an accretive structure. Beginning from an oblong single pen of roughly 12 by 18 feet and ranging as large as 18 by 24 feet, the houses grew in stages.

Additional modular pens were added as the need arose, and each room constituted a self-contained unit, almost in the manner of an apartment house. In some cases, internal connecting doors were absent. Perhaps the most typical enlargement never went beyond a second log room, but if expansion continued, it often occurred anomalistically and individualistically, defying attempts at a typology. Larger floorplan configurations that survive today vary from linear longhouses to T-, L-, and U-shaped plans.[17] These rambling, more sizable Hispanic dwellings served a well-established agrarian culture, in contrast to most Anglo-American housing, and multiple related families typically inhabited each. Log houses that remained small, with only one or two pens, bore the appellation *casita primitiva* (Fig. 2.17).[18]

Conceivably, the modularity of the New Mexican Hispanic house derives from the adobe or stone Spanish hacienda residence of Santa Fe and the central plateau of Mexico, which has independent rooms arranged around a central patio. The relatively uncommon U-shaped plan in highland New Mexico comes closest to duplicating the hacienda patio arrangement. Overall,

Figure 2.17. An Hispanic *casita primitiva* in highland New Mexico (research district 2). Most houses in this culture area developed into larger structures than this double-pen. Distinctive Hispanic features include exterior adobe plastering and the flat roof. The house now stands in the Old Cienega Village Museum near Santa Fe. (Photo by T.G.J., 1994.)

though, one is struck by how different the Hispanic log houses look when compared with traditional types in Mexico. They fit better a nonconformist model.

Ethnicity also helps explain high frequencies of atypical houses in the Peace River country, where a variety of non-Anglo groups are represented; in the Kenai Peninsula of Alaska, where native Americans perpetuate a hybrid Russo-Aleut log building culture; and in various other districts. The houses of the Aleuts are essentially Russian in design and concept (Fig. 2.18). In scattered ethnic colonies elsewhere in the West, one finds dwellings of Ukrainian, Finnish, and Scandinavian design.[19]

These ethnic-based dwellings remind us at the outset of our study that the West cannot be understood simply in Anglo-American terms. The region may, in the final analysis, owe much or part of its distinctiveness to the particular mix of cultures that assembled there. We must always be on the lookout for landscape traces of exotic groups and for evidence of cross-cultural influences. Even so, Anglo-Americans in the mountain West universally rejected these exotic house types, causing them to remain geographically encapsulated in ethnic pockets. Instead, borrow-

Figure 2.18. Russo-Aleut log house, Ninilchik, Kenai Peninsula, Alaska (research district 25). In architecture and carpentry, the house is Russian. (Photo by T.G.J., 1990.)

Figure 2.19. Polygonal Navajo *hogan* of notched logs, near Nazlini, Arizona, a classic example of ethnic influence in western log dwellings. Carpentry details reveal New Mexico Hispano influence, but the plan of the house is pure Navajo. (Photo 1994, courtesy of Stephen C. Jett.)

ing went the opposite direction, as many ethnic minorities, and in particular native Americans, adopted Euroamerican log construction and dwelling plans. Perhaps the most notable exception is provided by the Navajo, who accepted notched log construction from the New Mexico Hispanos but retained their traditional polygonal *hogan* dwellings (Fig. 2.19).[20]

Roadhouses

A typical landscape feature throughout much of the mountain West, and especially in northern districts, is the log roadhouse, or inn, which represents another regional dwelling type (Fig. 2.20). Trails were long and settlement sparse, making the common eastern practice of overnighting in homes less feasible. As a result, fully commercial roadhouses became the most common solution out West.

These inns follow no consistent architectural style. A few, in keeping with Midland eastern tradition, exhibit a "dogtrot" plan, as exemplified by the one at "mouth of Quartz" in Yukon Territory, in the year 1900.[21] However, most roadhouses are larger than the dogtrot, including some of the most sizable notched log buildings to be found anywhere in North America (Fig. 2.21).

Figure 2.20. Roadhouse of log construction, at Central, near Circle, Alaska (research district 22). Note the double-ridgepole construction. Road-houses come in many different architectural styles. (Photo by T.G.J., 1990.)

They include both gable-front and eave-front designs. One of the most impressive is the 108-Mile House, a National Trust complex of log buildings on the road to the Cariboo mines in central British Columbia. Sometimes these complexes bore the name "road ranches," providing fresh horses for traffic on the primitive highways of the West and fattened cattle for the mining markets, in addition to normal inn services.[22]

Heating the Dwelling

The hearth forms the nucleus of any dwelling, and especially so in cold climates. In the eastern United States, the most common heating device for log houses is a British-derived open-hearth fireplace, positioned centrally in a gable wall and often multiple in double-pen structures. A masonry or clay-lined wooden chimney extending above the roof ridge ventilates the fireplace (Fig. 2.22). Perhaps the most outstanding trait of this eastern device is the inherent inefficiency, as it allows most of the heat to escape unused up the chimney. In spite of this defect, the English fireplace and chimney easily conquered its chief rival, the Pennsylvania "Dutch" stove.

While the eastern fireplace and chimney reached the montane West, especially where crop farming became established and settlement began before 1860, they never achieved their former dominance.[23] The great majority of western log houses have neither open hearth nor chimney (see Figs. 2.3, 2.7, 2.8, and 2.10). Instead, the Pennsylvania Dutch stove, modified into the mass-produced and misnamed "Franklin" model, ventilated by a metal pipe occasionally encased in brick, heats the western log house. These cast-iron stoves became commercially available and affordable when railroads reached the West, after about 1870. They proved far more efficient, requiring less firewood and delivering more heat. Although the resin-rich coniferous wood of the western forests coated stovepipes with a flammable, dangerous soot, occasionally presenting the spectacle of flames erupting from the top of the stack, the iron stove easily won the West.

The highland Hispanos of New Mexico had a somewhat different traditional method of heating. Their hearth, an open fireplace, often hooded, or an earthen oven, lay in the corner of the log pen and smoke escaped through a small chimney. Some used ovens located outside, in the yard, for cooking. The Hispanos, too, adopted the cast-iron stove when it became available.[24]

The abandonment of traditional hearths by both Anglos and Hispanos does not, of course, result from western innovations. Instead, it came with the manufacturing age. The cast-iron stove reflects the general and rapid decline of folk culture evident throughout most of rural America by the last half of the nineteenth century. Settlement of most of the mountain West coin-

Figure 2.21. Gakona Lodge and Trading Post, an Alaskan log roadhouse complex near the junction of the Glenn and Richardson highways. In operation since 1905 along the trails that later evolved into highways, Gakona grew over the years. The present main structure dates from 1929. In 1977 the complex was entered in the National Register of Historic Places. (Photo by T.G.J., 1990.)

Figure 2.22. An English-plan cabin with typical eastern chimney, at Bingham Springs, Umatilla County, Oregon (research district 11). As in the East, the lower part of the pioneer chimney consists of notched logs lined on the inside with clay. The upper part is of sticks and clay. A raftered, shingled roof covers the cabin, and the round logs are V-notched at the corners. While this house and its chimney would be typical in, say, the Appalachians, it is exceptional in the West. (Photo courtesy of the Oregon Historical Society, Portland, neg. no. ORHI-36118.)

cided with that decline, and part of western cultural distinctiveness rests on the weakness of an underlying folk tradition.

The study of log dwellings has begun to reveal the cultural outlines of the West. Their principal message is that multiple Wests exist, that no monolithic, simplistic region in the Webbian or Turnerian sense awaits explanation out there, but that there is, instead, a bewildering mosaic of subregions. The West displays a vivid cultural geography unanticipated by the historians. Nor will explanations come easily, even on a subregional level, for the log cabin of the West has sent us contradictory messages of innovation and continuity, archaic and new, folk and popular, mainstream and ethnic. Things appear infernally complicated out there.

We must, therefore, push our inquiries into other topics. Dwellings constitute only one element in a diverse culture complex of log construction in the West, and they cannot convey the whole picture. In hopes of clarifying the picture, we turn next to a consideration of diverse types of log outbuildings.

Chapter Three

LOG OUTBUILDINGS

Log outbuildings in great diversity of both form and function appear on the landscape of the West, bearing more clues concerning the origins of the sectional culture. Most of these structures are directly connected to livelihood, to wresting a living from a difficult if beautiful land, and as a result they reveal adaptive strategies of land use perhaps even better than dwellings do. Barns of one kind or another account for the majority of western log outbuildings, and they possess considerable diagnostic potential. As one distinguished student of the West, J. B. Jackson, has said, "no feature of the landscape is more suggestive of the region's complex past than the type of barn or storage house."[1]

Implicit in Jackson's remark is the diagnostic potential of a comparison between western outbuildings and their counterparts in eastern North America. As was also true of dwellings, the folk barns of the East have been thoroughly and extensively studied, within the identical regional framework of Midland, Yankee, French Canada, and Tidewater South. We present in this chapter precisely such a comparison. In the process we can test Richard Francaviglia's belief that western barns "apparently originated in the eastern states," undergoing modifications "to meet the needs of western farmers and ranchers."[2]

Eastern Barns in the West

In the Midland East, four major log barn types dominated. The most common and simplest, a small or modest-sized *single-crib* structure, consists of one log unit with an entrance in a gabled end. It combines a maize granary below with an attic hayloft. In plan, this single-crib barn closely resembles the Anglo-Western cabin and the two likely bear a genetic kinship. The entrance to the maize granary is often merely a small hatch opening, and the hay loft either has a similar door or can be reached through an open gable.[3]

A log barn similar to the eastern single-crib represents the most common western mountain type (Figs. 1.2 and 3.1). Very widely distributed, the western gable-front single-crib accounts for about 22 percent of the 400-odd log barns we inspected in the West, nearly one-fourth of the total (Fig. 3.2). The hatch door of the East gives way to a full-sized entry in the great majority of western single-cribs, a hint that the resemblance is somewhat superficial. The lower level of the western barn rarely serves as a granary, reflecting the failure of maize to accompany Anglo-American pioneers into the mountain valleys. Often, in fact, the western single-cribs lack any internal division, and many are not really barns, in the proper sense of the word, serving instead a variety of functions, from toolshed to tack room. If a hayloft is present, then the lower level normally houses a horse stable. Some western single-cribs are obviously former dwellings demoted to the status of outbuildings. In spite of these functional differences, the fact remains that the most common type of eastern log outbuilding also occupies that position in the West, providing a strong message of continuity, mildly conditioned by readaptation.[4]

By contrast, other eastern barns largely failed in the West. The second most common log barn of the East is the *double-crib*, consisting of two log units facing a common wagon runway, all covered by a single span of roof. It duplicates the plan of the dogtrot house, to which the double-crib is undoubtedly related. Often the runway remains open to the air, but doors appear on many. Eastern double-cribs usually also combine lower-level granaries with attic haylofts, though stabling in one crib is not uncommon. The double-crib type, also called the "English barn," appears abundantly in both the Midland tradition and in the Yankee–Anglo-Canadian culture area, particularly Ontario.[5]

We found only twenty-three double-crib log barns in our western research districts, amounting to but 6 percent of the total (Figs. 3.2 and 3.3). In other areas we and other scholars came across an additional ten specimens.[6] A single district, highland New Mexico, accounts for fully 83 percent of the double-cribs noted in our sample, and their concentration in that largely Hispanic area calls into question the eastern affiliation of the montane western type. We deal with this issue later in the chapter. All western double-crib barns lacked maize granaries, and those built by Anglo-Americans generally served as hay sheds and stables. The eastern double-crib barn shared the fate of the dogtrot house in the West, achieving implantment but not noteworthy

Figure 3.1. Single-crib, gable-entrance barn, measuring about 16 by 20 feet, in the central Coastal Mountains of British Columbia (research district 18). Built by a preemption settler named Fred Hampton about 1912, the barn now stands in the Terrace Heritage Park. (Photo by T.G.J., 1990.)

acceptance. The same climatic unsuitability probably doomed both.

Similarly, several major types of larger eastern barns also failed in the West. The so-called *Pennsylvania* barn, linked culturally to the colonial German people of that state, consisted in its prototypical form of a double-crib log structure enlarged to two-story height. The upper level is accessible by an earthen bank or ramp at the rear, upslope eave side. On the opposite eave side, the upper level projects as a cantilevered forebay, providing the most distinctive feature of the type; the lower level is devoted to stabling, mainly of cattle, while the larger upper story includes grain and hay storage. The Pennsylvania bank barn became a common Midwestern type, known to many of the settlers of the mountain West. Even so, we know of only two specimens of the Pennsylvania bank barn in the entire West, neither of which involves log construction (Fig. 3.2).[7] Suited admirably for a Teutonic Corn Belt system based on grain cultivation and live-

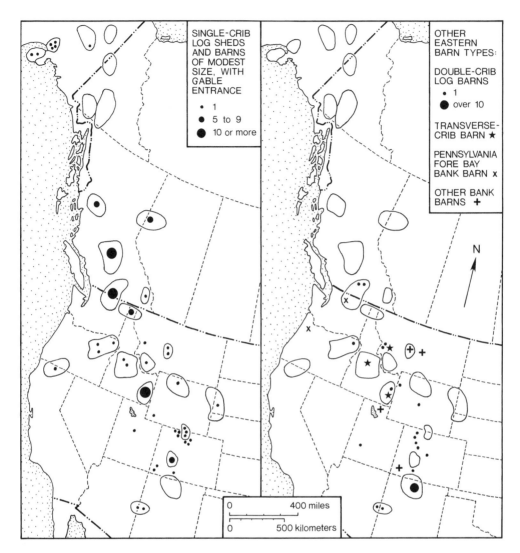

Figure 3.2. Distribution of eastern-derived barn types in the montane West.

stock fattening, the Pennsylvania bank barn proved unsuited to the agricultural strategies of the West. Indeed, bank barns of any description are very rare in the West.

The other major eastern type of log barn, the *transverse-crib*, has entries in both gable ends, a central through-passage connecting them, and multiple cribs lining each side. These four to ten log cribs are devoted to granaries, feed, storage, and stalls, while above them is a commodious hayloft. In many cases, each crib forms an independent log unit, but just as often the side walls of the barn consist of a single span of logs. Linked culturally to the Upland South, the transverse-crib serves well the mixed farming system of that region, including maize, lean cattle, and hogs. The failure of that mixed system, and particularly maize and hogs, to gain a foothold in the mountain West, pretty much excluded the transverse-crib barn from the region, even though Appalachian whites founded colonies here and there (Fig. 3.2).[8] For all practical purposes, the transverse-crib is absent from the West.

The Mountain Horse Barn

While the eastern transverse-crib is not a western type, many barns there outwardly resemble it. Notable among these is the mountain horse barn, a sizable gable-entrance log structure averaging about 27 by 40 feet (Fig. 3.4). One magnificent specimen at 108-Mile House in British Columbia measures an astounding 160 feet in depth, a span achieved by splicing logs. An entrance, positioned in the middle of the front gable end, leads into a central passage. Towering above the lower level is a huge raftered and gabled roof, either gambrel or of unbroken pitches, which permits a very sizable hayloft, accessible through a door in the gable.[9] In these respects, the mountain horse barn closely resembles the transverse-crib.

Closer inspection reveals it to be a basically different type. For one thing, the mountain horse barn is a single-crib structure, whereas the eastern transverse-crib consists of multiple cribs

Figure 3.3. Double-crib barn with open wagon runway near Swan Valley, Bonneville County, Idaho (research district 7). This eastern type is rare in the West. Note the board roof. (Photo by T.G.J., 1987.)

Figure 3.4. Mountain saddle horse barn at 108-Mile House, British Columbia (research district 16). It has a huge hayloft, raftered roof, and a side shed room. (Photo by J.T.K., 1990.)

joined by one overarching roof span. More important, fundamental differences in floorplan exist. More often than not, the mountain horse barn lacks a rear door, and the central passage does not function as a wagon runway. In the place of maize granaries are box stalls with doors or slat gates, open tie stalls with mangers, and a tack room, all arranged in a U shape around the sides and rear of the barn in those cases where a rear door is not present. Sometimes, small windows in the side walls illuminate the lower level (Fig. 3.5). Usually the internal partitioning of the mountain horse barn is achieved by board walls, though some have log dividers notched into the side walls (Fig. 3.6). The hayloft, which often extends well down into the part of the barn enclosed by log walls, typically rests upon large morticed joists.

This distinctive barn, linked to the highland cattle ranching industry, serves to stable a remuda of saddle horses in winter, to store enough hay to feed them in that season, and to keep riding equipment. Range cattle require no cold-season stall feeding, but saddle horses receive high-quality care. The largest, most elongated mountain horse barns served stagecoach stops at roadhouses or, in one case at least, stabled mules used to haul coal to cooking ovens.[10]

Mountain horse barns appear frequently throughout the high-valley ranching country, from central Colorado to southern British Columbia (Fig. 3.7). We inspected forty-six such structures in our various research districts, about 12 percent of all western log barns, and encountered many others elsewhere. The origin of this barn type is almost certainly western and most likely occurred in southwestern Montana, where the earliest high-valley cattle ranching complex developed during the 1860s.

Figure 3.5. Western horse barn with huge hayloft, supported by morticed joists, and a fine raftered, shingled roof, Okanogan County, Washington (research district 15). Note the side windows on the stall level. (Photo by T.G.J., 1990.)

Figure 3.6. A smaller than average western horse barn, Boundary County, Idaho (research district 13). Note the log partition walls and hay doors. The front entrance does not accommodate wagons, unlike the barn shown in Figure 3.4. (Photo by T.G.J., 1987.)

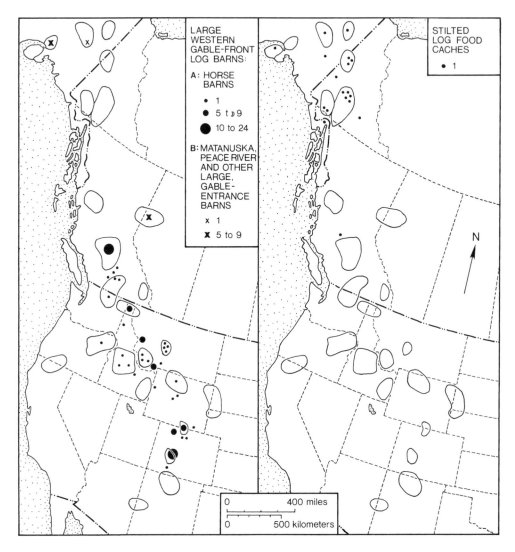

Figure 3.7. Distribution of large western gable-front barns and stilted food caches.

That region lies almost central in the distribution of the mountain horse barn.

Perhaps the eastern transverse-crib barn exerted some superficial influence in the evolution of this western mountain type. Certainly the outward configuration is very similar. The two barn types served, however, fundamentally different agricultural systems, as is revealed in their internal arrangement. Moreover, upland southerners, including Texans, remained rare in the districts where the mountain horse barn occurs. Instead, the founders of high-valley ranching came from the Midwest, where the transverse-crib barn was never common.[11]

Matanuska and Peace River Barns

Two other regions of the West have large, gable-entry barns out-wardly similar to both the mountain horse barn and the trans-verse-crib, but probably unrelated to either. One of these, the Matanuska barn of the Anchorage hinterland in southern Alaska, is distinguished by its square plan, almost invariably measuring 32 feet on both sides and ends. Also, the wall logs are spiked rather than notched at the corners. Carpenters hired by the United States government built these barns to specifications and blueprints in the 1930s, explaining their uniformity of dimensions and shoddy workmanship.[12] The Matanuska barn is de-signed to serve a dairy economy, combining milk cow stalls with hay storage (Fig. 3.7).

Very similar in appearance, though older and built of notched logs, are the barns of the Peace River country along the Al-berta–British Columbia border (Figs. 3.7, 3.8). This region, like the Matanuska Valley, experienced one of the last agricultural colonizations in the North American West. More elongated than the Matanuska type, the Peace River barn stables both cattle and horses. Wheat granaries occupy multiple, freestanding single-crib structures in the Peace River country. The highly diverse ethnic population of the area makes it difficult to decipher the cultural history of the Peace River barn, but it seems to have no genetic tie to either the western horse barn or the transverse-crib. It is, in any case, not properly a mountain western barn, but instead belongs more to the Canadian prairie provinces.

Western Eave-Entry Barns

Several other distinctively western types of log barn appear in the cultural landscape. Included are various eave-entry types that rarely, if ever, occur in the East. The most common of these is a single-crib barn or shed, ranging in size from modest and low to more sizable, two-level structures (Fig. 3.9). Like the gable-entrance mountain barn, this type most often combines horse stalls, tack room, and hayloft. The two appear in about the same geographical area (Fig. 3.10). We found fifty-nine such barns, accounting for about one in every seven in the West. No eastern prototype is evident, and the eave-entry single-crib likely represents a western innovation. In fact, it offers one of the few examples of eave-entry log structures that achieved notable ac-ceptance in the West.[13]

Closely related but rather uncommon is a double-crib type, in which the two log units abut one another (Fig. 3.11). These ap-parently developed as enlargements of the more widespread, smaller type. Even less common is another double-crib type fea-turing a very narrow passage between the two log units (Fig.

Figure 3.8. Peace River country log barn, DeBolt, Alberta (research district 17). Note the bellcast gambrel roof, which suggests ethnic influence, possibly Ukrainian or French-Canadian, and the chinkless full-dovetail notching (Photo by T.G.J., 1990.)

Figure 3.9. Single-crib, eave-entry horse barn of modest size in southern British Columbia (research district 15). Note the morticed joists supporting the hay loft and the board roof. (Photo by J.T.K., 1991.)

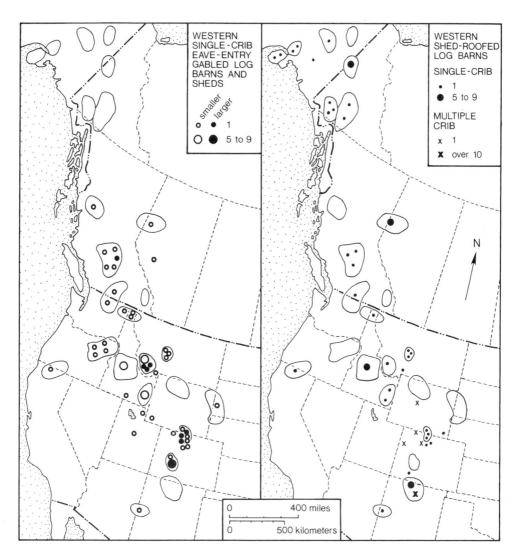

Figure 3.10. Distribution of western eave-entry log barns, including both gabled and shed-roof types.

3.12). The narrowness of the passage disqualifies this barn as an eastern double-crib, as does the appearance of a second entry on the gable end.

The numerous shanty-type structures distinguished by a single-pitch shed roof and a door in the taller, front-facing wall are best placed in the category of eave-entry barns (Fig. 3.13). Western cribs of this sort accounted for about 15 percent of the barns in our survey and served many functions. In the Rocky Mountains, Anglo-Americans employ them as chicken coops or as sheds for haying equipment, among other uses. Some even served as bunkhouses. Farther north, such sheds often stand ad-

Figure 3.11. An abutted double-crib barn with large hayloft south of Cascade in Valley County, Idaho (research district 10). Such a barn would look altogether out of place in the East and helps establish a distinctive western cultural identity. (Photo by T.G.J., 1987.)

Figure 3.12. Distinctive western double-crib barn, differing from the eastern type by virtue of its narrow central passage and the presence of another door on the gable end. This horse barn stands near Columbia Lake, British Columbia (research district 14). Too few such barns occur to justify a separate type. (Photo by T.G.J., 1987.)

Figure 3.13. Single-crib, shed-roofed log storage building, a type common in the mountain West, located at Gakona Road-house, near Glennallen, Alaska. (Photo by J.T.K., 1990.)

jacent to roadhouses, where they were used for food storage. Most of these shanties are of single-crib size, but some include two or even more, especially in Hispanic New Mexico, as will be described later (Fig. 3.14).[14]

While these shed-roofed log structures could be of western origin, the East offers abundant examples of similar outbuildings. In those parts of southern Appalachia where traces of the folk culture remain strongest, shed-roofed chicken coops or hog shelters often survive to the present day. Log dwellings of shanty design also once existed in parts of the East, especially Ontario.[15] These common shed-roofed outbuildings likely represent an archaic eastern form that has achieved greater survival in the West, and the widespread distribution of the shed-roofed crib in western areas suggests diffusion from the East (Fig. 3.10). The same may be true of the shed-roofed western livestock shelters, most often sheep folds, consisting of only three log walls, with the remaining, leeward side left completely open. They bear a strong resemblance to the "three-faced camps" built as crude log shelters by hunters in the frontier East and long since vanished there.[16]

A great many other western barns simply defy a typology, as we also found true of dwellings. They exhibit diverse, unique plans. We placed fully 125 barns, about 31 percent of the total, in that category. A disproportionate number of these lie in Hispanic New Mexico, and we need now to turn our attention to that distinctive ethnic enclave.[17]

Hispano Log Outbuildings

Any observant visitor to the Hispanic highlands of northern New Mexico will be struck by the abundance of log outbuildings there, by the strikingly different appearance most of them ex-

Figure 3.14. Multipen, shed-roofed barn, Garfield County, western Colorado, which apparently grew in stages. (Photo by T.G.J., 1987.)

hibit in comparison to those found in Anglo-American areas of the West, and by a diversity of forms so profound as largely to frustrate most attempts at classification into types. Inquiries concerning the generic names of these structures evoke confusing and often contradictory replies. For example, residents of the mountain village of Truchas in Río Arriba County used four different Spanish generic words to name the very same log barn! In form, "no two look alike" and field measurements failed to disclose many uniformities.[18]

These apologies aside, we can impose some limited typology on the seeming Hispanic chaos. At the risk of oversimplification, we can say that a *tasolera* is a shed-roofed barn containing a log-walled shelter for horses, burros, cattle, sheep, or goats at the lower level and a hayloft beneath the single-pitched roof (Fig. 3.15). These range from single-cribs of modest proportions to multicrib barns up to 65 feet long. Local tradition suggests a recent, indigenous origin of the shed-roofed *tasolera*, apparently as recently as the 1920s. While it conceivably could be a borrowing from Anglo-Americans, that seems unlikely (see Fig. 3.14). Shed-roofed barns became in a short span of time far more common among Hispanos than Anglos (Fig. 3.10). In addition, many *tasoleras* differ from comparable Anglo structures by virtue of a stilted roof support that enlarges the hayloft and produces a distinctive appearance (Fig. 3.16). Anglo shed-roofed barns most

Figure 3.15. Elongated, multicrib *tasolera* (also sometimes called a *caballerisa*), joining a lower-level stable and an attic hayloft beneath a shed roof, in Truchas, Río Arriba County, New Mexico (research district 2). (Photo by C.F.G., 1966.)

Figure 3.16. Stilted-roof, multicrib *tasolera*, a type of fairly recent origin, in Córdova, Río Arriba County, New Mexico (research district 2). Livestock are housed in the lower level. (Photo by C.F.G., 1968.)

often lack haylofts altogether. Also arguing against diffusion from the Anglos is the fact that many *tasolera* cribs are entered from the side walls rather than the front. The Nahuatl-derived name remains a mystery to us.

Similarly, a double-crib log barn with open central runway very much like those of the Anglo-American type appears among the Hispanos (Fig. 3.17). These contain stalls on the lower level and haylofts above, with raftered, gabled board roofs. They, like the shed-roofed *tasoleras*, appeared late in the Hispanic highland culture. The first double-crib was erected about 1907, also in the Truchas area, conceivably as another borrowing from Anglo-Americans. More likely, it represents an independent, indigenous invention, as was suggested in our interviews with local folk builders. Hispanos use no consistent generic name for these double-cribs, though *caballerisa* is often heard. Certain examples, sometimes called *tapeistas*, have a distinctly ethnic appearance, with flat roofs atop which hay and maize are dried (Fig. 3.18). Often the dimensions of these *tapeistas* depart radically from any possible Anglo prototype, with individual log cribs measuring 11 to 14 feet in width and up to 30 feet in depth. In other cases, multiple double-crib barns form odd, abutted juxtapositions never found among Anglos (Fig. 3.19).

Figure 3.17. Hispanic double-crib barn with gabled board roof, reputedly the oldest example and prototype in research district 2, dating from about 1907, Truchas, Río Arriba County, New Mexico. (Photo by C.F.G., 1966.)

Figure 3.18. Hispanic flat-roofed double-crib barn, now at Old Cienega Village Museum, Santa Fe (research district 2). The flat roof serves as a *tapeista*, or platform to dry maize and hay. The dimensions of the cribs are 11 by 30 and 14 by 30 feet, a highly distinctive plan. (Photo by T.G.J., 1994)

Figure 3.19. Distinctive Hispanic lumping together of a double-crib barn and other log structures, Truchas, Río Arriba County, New Mexico (research district 2). (Photo by C.F.G., 1969.)

Many remaining Hispano barns resist classification. They may take on a linear aspect, with as many as five or six linked cribs devoted to diverse functions such as stabling, hay storage, tack room, hog pen, granaries, sheep folds, and goat sheds. We might best call these Hispanic *long barns*, and they form a parallel development to some dwelling expansions. In other cases mazelike complexes of corrals and log cribs simply grew like Topsy. (Fig. 3. 20).

Often the Hispano farmstead takes on a strewn appearance, with a variety of special purpose outbuildings supplementing the *tapeista* and *tasolera*. While most often single-cribs, these structures vary so much in form that they are better grouped according to function. A *barbacoa*, in spite of its name, is a maize granary consisting of a flat platform roof for the stalks and a small log crib beneath in which the corn is stored (Fig. 3.21). Walls are often chinked or plastered to minimize the loss of grain to rodents. A *cochera*, literally a coach or wagon house, can most readily be identified by a large door or opening in one of the structure's narrow ends. The *dispensa*, a shelter in which miscellaneous nonperishable household items such as clothing are stored, is normally contiguous or close to the house, and has a locked door (Fig. 3.22). The *fuerte* is a small storage stronghouse for harnesses, tools, small equipment, and a variety of other items (Fig. 3.23). A

Figure 3.20. A complicated cluster of log outbuildings in an Hispanic farmstead, Córdova, Río Arriba County, New Mexico (research district 2). (Photo by C.F.G., 1971.)

Figure 3.21. A *barbacoa*, or corncrib, in Hispanic highland New Mexico (research district 2). Note the stone foundation and the platform where corn-stalks or hay are stacked above the storage room. (Photo by C.F.G., 1966.)

unique two-level *fuerte* appears in a limited distribution east of the Sangre de Cristo divide. Its *alto*, or attic, accessible by ladder through a door in one of the gable ends, is used for storing items of relatively low value or infrequent use. The lower room serves for general storage (Fig. 3.24). A *troja* (or *soterrano*) provides a pantry for foodstuffs. The perishable nature of the holdings re-quires some protection against freezing temperatures and ro-dents, and the structures have chinking in the log interstices, adobe plastering, a lock, and interior shelves and cabinets. These diverse small outbuildings often confound our Anglo-based ef-forts to classify them as gable-front or eave-front because of their flat roofs.[19]

With the possible exception of the *troja*, few if any of the di-verse highland Hispanic barns and other outbuildings seem to be derived from types in old Mexico. They are probably best re-garded as largely indigenous, as a diverse and individualistic re-sponse to a markedly colder habitat than that of the Mexican central plateau.

Other Western Log Structures

Western cultural landscapes contain numerous other kinds of log structures, such as churches, schools, smithies, courthouses, forts, gristmills, jails, dams, bridge abutments, flume supports, saunas, and assorted mine buildings.[20] We will consider only a few types here, since the list is almost endless.

No log structure is more distinctive or potentially revealing than the stilted *food cache* of the far North (Figs. 3.7 and 3.25).[21] So common in Alaska and the Yukon territory as to have become

Figure 3.22. A log *dispensa*, for storage of household items, in Truchas, Río Arriba County, New Mexico (research district 2). Such outbuildings are common throughout the Hispanic highlands. (Photo by C.F.G., 1966.)

a regional symbol and icon, the food cache, a diminutive frontgable crib, stands 6 to 12 feet above ground level on four posts. Accessible by ladder, it provides a place to store meat safe from scavengers. Two possible explanations exist concerning its origin. Virtually identical stilted log caches exist among the Sami people of northern Scandinavia and Russia's Kola Peninsula.[22] Conceivably, Russian fur traders brought the log cache eastward across Siberia and into Alaska. These structures, however, do not appear in the Russianized Aleut settlements, nor is Russian-style carpentry employed in their construction. More likely, the stilted food cache of nonlog construction existed traditionally among native boreal hunting peoples in an unbroken forest belt from Scandinavia to northwestern North America. As these various tribes acquired steel axes and notched log construction from Caucasian intruders, each adapted the new carpentry for their caches. In turn, the concept of an elevated cache diffused in the opposite direction, and Anglo-Americans adopted it from the Dene tribes of interior Alaska and the Yukon. Still today, stilted caches remain most common in Indian villages (Fig. 3.26).

Figure 3.23. A log *fuerte*, a shed for storing farm tools and equipment, at Pecos, San Miguel County, in Hispanic New Mexico (research district 2). (Photo by C.F.G., 1966.)

Figure 3.24. Unusual two-level *fuerte*, at Ledoux, Mora County, New Mexico (research district 2). The roof is exceptionally steep, even by local standards. (Photo by C.F.G., 1966.)

Diffusion of the food cache from Amerindian to white settlers suggests a previously neglected explanation for some aspects of western cultural distinctiveness, one unanticipated by any historian of the West, new or old, or by anyone else for that matter. Borrowing from the culturally distinctive native people may help explain western sectionalism. We must look for more causes than merely diffusion from the East and western innovation. The cultural landscape has, in the humble food cache, offered us an explanatory clue that no archive could contain.

Gristmills are very rare in mountain western landscapes, a reflection of the sectional unimportance of grain cultivation. The major exception is agrarian Hispanic New Mexico, where, at least as early as 1756, small water-powered mills housed in flat-roofed, single-pen log structures came into use (Fig. 3.27).[23] Their mill

Figure 3.25. Stilted log food cache, Porcupine, Alaska, near the Chilcat River west of Haines in the panhandle region. Food, especially meat, remains safe in such caches and the structure functions as a natural "refrigerator" for nine months of the year. (Photo by T.G.J., 1990.)

Figure 3.26. Stilted log food cache, Kluckshu Indian village, Yukon Territory (research district 20). The basic concept of this structure is likely derived from Amerindians of the northern forests. (Photo by T.G.J., 1990.)

wheels move counterclockwise in a horizontal plane, as in Mexico, southern Europe, and the Middle East, presenting a fundamental contrast to Anglo-American mill technology. Construction is generally crude, a clear indication that the housing serves mainly as a shell to protect the mill, grain, and flour from wind, rain, and pilferage. The comfort of the miller was of little consequence, since ice prevented milling during subfreezing weather. Floorplans are normally square, or nearly so, with outer dimensions varying from 9 to 15 feet. Ceilings stand quite low, measuring 7 to 8 feet in height. Floors consist of planks, covered with adobe to prevent the loss of grain through cracks. Low doorways, a trait common to the Hispanic tradition, require one to stoop upon entering. Windows are small openings, many only wide chinks between logs or narrow incisions on adjacent logs.

The Hispano mills lie astride small diversion ditches, lacking elaborate dams or millponds (Fig. 3.28). Ditches scarcely a foot wide and less than a foot deep provide enough flow to turn the horizontal mill wheel. Two or more of the small mills sometimes

Figure 3.27. Flat-roofed Hispanic gristmill at Taos, New Mexico (research district 2). (Photo courtesy of the Museum of New Mexico, Santa Fe, neg. no. 11481.)

Figure 3.28. "Old Mill at Chamita," Hispanic highlands of New Mexico (research district 2). (Photo by William Henry Jackson, ca. 1890, courtesy of the Denver Public Library Western Collection, no. 2604.)

occupy the same ditch. The flow of water to the mill wheel is always controlled. Because the wheel, axle, and "runner," or upper millstone, form a single, fused unit, the mill stops only when the water flow is diverted. Grain feeds from a suspended hopper into the "eye" of the runner, which rotates upon a stationary lower stone. The lack of an intermediate gear mechanism means that runner and waterwheel rotate at the same speed, all resting upon a single-point bearing at the bottom. In all but their log construction, these mills find a prototype in Spain. Regrettably, none any longer function.

The Hispano highlanders of New Mexico should by now have our attention. Their houses, barns, sheds, and mills all label this corner of the West as *highly* distinctive. We must, accordingly, seek diffusion not just from the Anglo-American East, but also the Mexican South and Iberian Europe. Ethnicity must be taken into account if we are to comprehend the West, and not just American Indian ethnicity.

We learned a fair amount from the diverse types of log outbuildings in the mountain West. They reminded us to pay attention to cultural mixing with Amerindians, to respect the strength of innovation and individualism, and to discard a wholly Anglocentric viewpoint. Perhaps even greater diagnostic potential lies in the details of notched log carpentry, a subject well studied in the East but not previously investigated in any comprehensive way in the western mountains. Chapter 4 is devoted to that promising topic.

Chapter Four   # Log Carpentry
Traditions

The various dwellings, barns, and other log buildings that we
have described exhibit a number of different folk carpentry methods, including log-shaping, notch-cutting, and the erection of
roofs. We found no fewer than eight distinct log carpentry traditions in montane western North America. The most common
among these is an Anglo-American Midland type, borne westward by the cultural group most closely linked to forest colonization, log cabins, and the backwoods frontier.[1] A second distinctive
eastern Anglo-American log building culture complex arose in
Yankee New England, most closely linked to the Piscataqua Valley on the Maine–New Hampshire border and to military blockhouse construction. This Yankee complex later influenced Ontario, especially the southern rim of the Canadian Shield, where it
finds greatest survival today. We will, accordingly, henceforth refer to it as the Anglo-Canadian tradition.[2]

Differing in basic ways from both Midland and Anglo-Canadian construction, the log carpentry traditions of the Russians,
French Canadians, and Mexicans reached the West well ahead of
the English-speaking settlers. Notched-log buildings existed in
both old and New Mexico by the middle of the eighteenth century and had probably been introduced by German priests or
mining engineers.[3] So imperfect and superficial is knowledge
concerning the culture of the West that a distinguished Smith-
sonian scholar not so long ago attributed notched-log construction among New Mexico Hispanos to diffusion from Anglo-
Americans.[4]

Complicating the picture still more, Scandinavians, Finns, and
Ukrainians later introduced their own log carpentry methods in

a scattering of ethnic colonies and individual homesteads in the West (Fig. 1.3).[5] While each of them remained ethnic-specific, failing to achieve diffusion to other groups, they add to the complexity of western identity. Carefully analyzed, log carpentry will reveal a cultural confluence and mixing that lies at the very root of western character.

Shaping the Log

Perhaps the most basic question addressed by the carpenter concerns the shaping of the log. The simplest and crudest solution is to leave the timber in its natural round shape, either peeled or with the bark intact (Fig. 4.1; see also Figs. 2.7, 2.17, and 3.26). Fully two-thirds of all the buildings in our twenty-five research districts consisted of round logs, and in three areas the proportion exceeded 80 percent (Fig. 4.2). The resultant rusticity of appearance contributes to the image of the West. Both Midland Anglo-Americans and Hispanos had a heritage of building with round logs, but in the eastern United States and Mexico the proportion of log structures displaying such carpentry is much lower than in the West. In the Midland Anglo tradition, round-log construction was more closely linked to the frontier, or "cabin stage," and in postpioneer times such crude carpentry declined, becoming relegated mainly to barns, sheds, and other outbuildings.[6] The prevalence of round-log construction in the West provides vivid evidence that an archaic eastern practice remains the most common form in the mountains.

Conversely, the most typical Midland practice, involving flattening the sides of the logs with ax and adz to produce hewn exterior and interior walls, remained far less common in the West (Fig. 4.3). Finer buildings, such as stores, churches, and inns, more often have hewn walls than do dwellings and outbuildings. Most commonly, hewn-log construction appears among ethnic minorities, especially Scandinavians, Finns, and Russianized Aleuts.[7] While no research district has hewn-log carpentry as the most common type, the Russo-Aleut western half of district 25 contains almost exclusively hewn structures. In some Anglo-American log houses, the builders hewed only the interior walls, performing this task after the dwelling was raised, a practice also documented for the frontier East.[8]

A third, intermediate method of western log-shaping involves hewing only the ends, near the notches, leaving the remaining, greater portion of the timber round (Fig. 4.4). This method accounts for 15 percent of all the log buildings in our research districts. Both in terms of actual numbers and in proportion of the total, end-hewing occurs most commonly north of the 49th parallel, the main United States–Canada boundary. Almost two-thirds of all examples stand north of that border, and the only

Figure 4.1. Round-log construction with two-sided saddle-notching, near Jackson, Montana (research district 9). (Photo by J.T.K., 1989.)

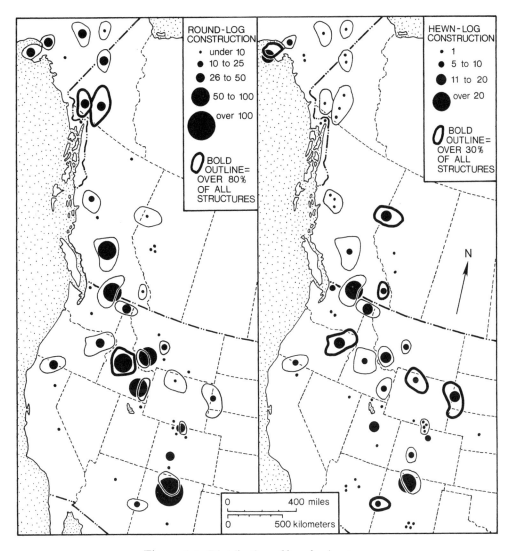

Figure 4.2. Distribution of log shaping types.

two research districts where end-only hewing was the most
common log-shaping technique lie wholly or partially in Canada
(Fig. 4.5). Even so, the technique often appears in Montana and
Idaho, south of the border.[9] While end-hewing does occur occasionally in the Midland Anglo-American carpentry tradition of
the East, it is so atypical there that the occurrence in the West
demands another explanation.

In part, ethnic influence is at work. Twelve of the thirteen examples of end-hewing in the Kenai Peninsula of Alaska occur on
Russo-Aleut structures, and we should not doubt that the Russians, as well as Finns, introduced the technique, given its widespread east Baltic occurrence. More important, end-hewing be-

Figure 4.3. Hewn, chinked, V-notched logs, Gunnison County, Colorado (research district 3). In each of these three respects, the carpentry is essentially identical to the most common eastern method of log
preparation, except that the logs seem to have been milled rather than
hewn with ax and adz. (Photo by J.T.K., 1989.)

longed to the eastern Anglo-Canadian tradition. Fully 45 percent
of all log buildings in the southern Shield area of Ontario exhibit
such carpentry. Present in the colonial Yankee military building
tradition, end-hewing became even more common in Ontario as a
result of a nineteenth-century law exempting round-log structures from taxation. With the exception of Russo-Aleut coastal
Alaska and a scattering of Finnish colonies, end-hewing in the
West provides an index of Anglo-Canadian and, more specifi-
cally, Ontarian influence. It warns us not to assume that all log
carpentry diffusing from the English-speaking East derives from
the Pennsylvanian Midland culture complex.[10]

Similarly, the virtual absence in the mountain West of Midland half-log construction, in which builders split round timbers lengthwise before using them in the walls, suggests less than a pervasive influence emanating from the most important eastern log carpentry complex. We found only one half-log structure in the entire West, in spite of the fact that the technique is common in eastern areas of pine-log usage.[11]

Lengthwise Fitting

The easiest method of fitting logs into a wall involves leaving a crack, or *chink*, between them, so that the timbers touch only at the corners of the building (Fig. 4.6). If the structure needs ventilation, as in the case of hay barns, the chinks remain open, but the normal practice otherwise is to stop and daub the cracks to achieve a weathertight wall. Chink construction is the almost universal Midland Anglo-American method and well-nigh diagnostic of that tradition. In the mountain West, most districts exhibit an overwhelming dominance of this simple construction

Figure 4.4. End-only hewing with full-dovetail notching, De Bolt, Alberta, in the Peace River country (research district 17). Both this shaping method and the notch type reflect Ontarian Anglo-Canadian carpentry influence. (Photo by T.G.J., 1990.)

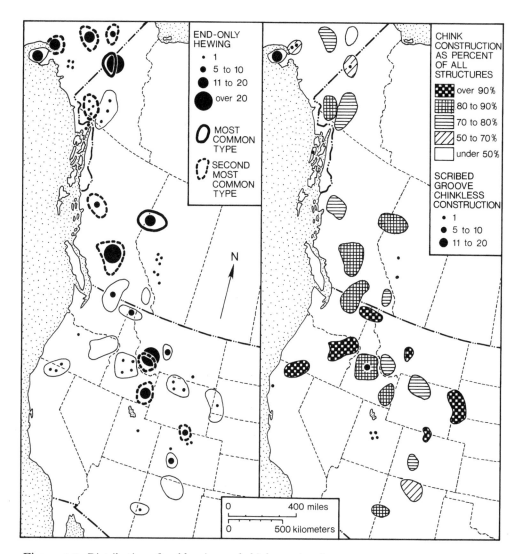

Figure 4.5. Distribution of end hewing and chink construction.

technique (Fig. 4.5), with proportions usually greater than 75 percent. Most western chinks are narrower than those observed in the Midland East, probably because straighter timbers were available. A variety of chinking material appears in western log buildings, from boards, slats, rails, and willow poles to torn army blankets. As in the East, the daubing consists of clay or mud mixed with grass or straw.

Only in Hispanic New Mexico and the far North is chink construction less common. When hewing logs, Hispanos usually flattened all four sides, allowing a tight lengthwise fit. This practice derived from the log carpentry tradition of Mexico (Fig. 4.7). In the Southwest, chinkless construction is virtually diagnostic of Hispanic influence. Russians introduced a technique in

Figure 4.7. Hispanic chinkless, hewn-log carpentry with oversided vertical single-notching in highland New Mexico (research district 2). In terms of both notch and lack of chinks, this structure reflects the carpentry influence of old Mexico. It stands now in Old Cienega Village, Santa Fe. (Photo by T.G.J., 1994.)

Figure 4.6. Chinked log construction, Big Hole Valley, Montana (research district 9). Note also the end-hewing and half-dovetail notching. (Photo by J.T.K., 1989.)

the northern Pacific region of the West much better suited to extreme cold conditions. They cut a broad concave groove lengthwise on the underside of each log, allowing a tight "scribed fit" atop the natural rounded surface of the log below (Figs. 4.5 and 4.8). A compacted layer of moss rests between the logs, serving as insulation. Russo-Aleut buildings on the Kenai Peninsula display this method, as do smaller numbers of Russo-Dene structures north of Anchorage. Finns, Swedes, and Norwegians reintroduced this same northern European scribed technique into various ethnic pockets of the montane West.[12] Some Anglo-Americans later adopted grooved-log fitting in Alaska, the Yukon, and elsewhere, usually degrading it into less-demanding flat hewing of the top and bottom of each log. In the "log cabin revival" stage, beginning about 1970, the scribed fit achieved wide-spread western usage.[13]

Certain other groups also built chinkless log structures in the West, often involving excellent carpentry (Fig. 4.9). The French-Canadian tradition usually involved tight-fit logs, and four-sided hewing appeared often in Anglo-Canadian carpentry as well (Fig. 4.10).[14] Overall, chinkless log construction of diverse provenance occurs far more frequently in the western mountains than in the Midland Anglo-American East, reminding us once again of the cultural complexity of the West.

Corner Timbering

The key to any log structure, lending lateral strength and bearing much or all of the weight, rests in the corners. Perhaps no other aspect of log carpentry possesses as much diagnostic potential, since the different cornering methods often derive from specific source regions and individual cultural groups. In the great majority of western log buildings, 97 percent to be exact, timbers are *notched* together at the corners. Only three notch types—*square*, *V*, and *saddle*—account for nearly two-thirds of all western log buildings.[15]

Square-notching alone accounts for more than a quarter of all structures and is the most common type in eleven different western districts, from the Mogollon Rim of Arizona to the Mat-Su Valley of Alaska (Fig. 4.11). This type, also known in the West as the "tenon," "tennant," "lapped," or "box" corner, is easily formed by sawing out a small block from the top and bottom, at the butt end of each timber (Fig. 4.12). Equally suitable for round or hewn logs, the square notch has flat surfaces that do not provide a locking joint, and as a result spiking is necessary. Ease of construction no doubt helps explain the popularity of this notch type.[16]

The square notch occurs in the Midland Anglo-American tradition of the East, though not as commonly as in the West. It appears most often in areas peripheral to the Pennsylvanian-Upland Southern core of Midland log construction, such as the inner coastal plain of the South and Ontario. In that context, its importance in the West, another periphery, did not surprise us. Still, the explanation is not that simple. Square-notching also typified the Anglo-Canadian log culture complex of Ontario and its parent Yankee–garrison house tradition in colonial New England. In at least one documented instance, square-notching achieved direct diffusion from New England to the West.[17] Moreover, Russians also introduced square-notching into Alaska and California. The widespread occurrence of western square-notching reflects three different sources and is not diagnostic of Midland influence.

Figure 4.8. Scribed, grooved lengthwise fit on chinkless log wall in the Finnish ethnic community of Long Valley, in Valley County, Idaho (research district 10). Introduced by Russians and various northern European groups, grooving spread belatedly to some Anglo-Americans, mainly in Alaska. (Photo by T.G.J., 1991.)

Figure 4.9. Chinkless, milled, full-dovetailed log construction, Fort Steele, British Columbia (research district 14). The structure stands in the Fort Steele Heritage Park. Ontarian Anglo-Canadian influence is evident. (Photo by T.G.J., 1987.)

Figure 4.10. French-Canadian corner- and medial-posted log construction, Fort Vancouver, Washington. (Photo by T.G.J., 1987.)

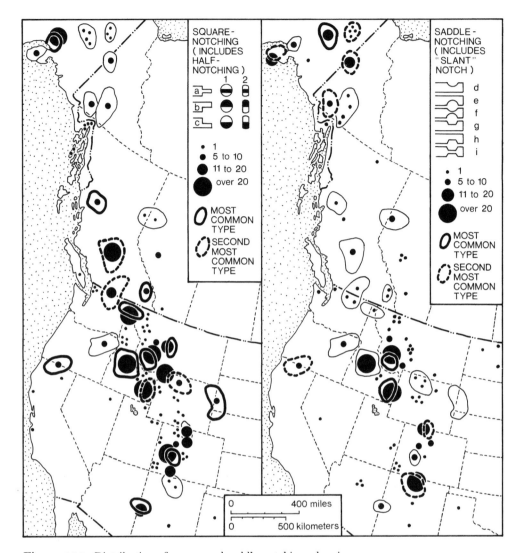

Figure 4.11. Distribution of square and saddle notching, showing subtypes. KEY: a = regular square-notch; b = undersided half-notch; c = oversided half-notch; d = oversided regular saddle notch; e = undersided regular saddle notch; f = 2-sided regular saddle notch; g = oversided slant notch; h = undersided slant notch; i = 2-sided slant notch; 1 = round-log; 2 = hewn.

Saddle- and V-Notching

One-fifth of western log buildings displays the *V-notch*, a type named for the internal shape of the joint, which forms an inverted V (Figs. 4.3 and 4.13; see also Figs. 2.9 and 2.22). Also called the peak-and-keeper or saddle-and-rider notch by local people, this type of cornering occurs throughout the mountain West and represents the most common notch in six research districts in the central section, excluding Mormon areas (Fig. 4.14).

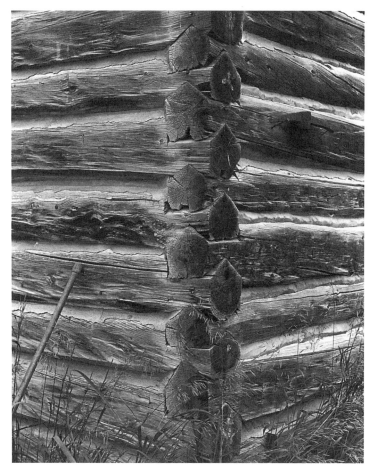

Figure 4.13. Hewn, uncrowned V-notching of the Pennsylvanian type, Virginia City, Montana. Any Midland carpenter back East would find this familiar. (Photo by T.G.J., 1987.)

Figure 4.12. Square notching on round logs, Blue Mountains, Oregon (research district 11). This carpentry combination is very common in the West. (Photo by T.G.J., 1991.)

The V-notch possesses diagnostic capability, for it is unique to the Midland Anglo-American carpentry tradition, appearing in no other North American log building complex. The widespread distribuiion and importance of V-notching in the West provides impressive evidence of the cultural reach and importance of Pennsylvania Extended.[18]

Western V-notching can instruct us still further concerning cultural history. In its most common eastern form, the V-shape of the notch carries to the butt end of the log, but in a very rare, demonstrably archaic subtype, the log resumes its natural shape outside the notch, producing a crowned end (Fig. 4.15). Perhaps one in a hundred eastern examples of V-notching exhibits crowning, even though the type originally arrived in the colonial Delaware Valley from Sweden in the crowned form. In the mountain West well over half of all specimens bear crowns (Fig. 4.14).

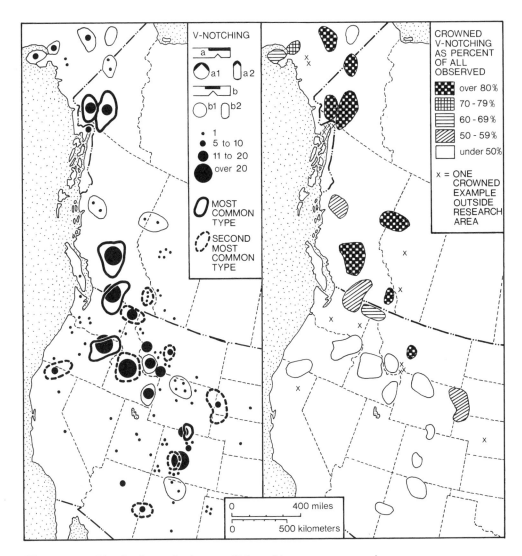

Figure 4.14. Distribution and subtypes of V-notching: a = uncrowned;
b = crowned; 1 = round-log; 2 = hewn.

In northern areas, crowned V-notching regularly exceeds 80 percent of those observed, and all eighteen specimens found in the
Kluane country in Yukon Territory display this characteristic.[19]
We can conclude that western V-notching not only demonstrates
important links to Midland Anglo-America, but also reflects a
dominantly archaic form of Midland culture. This detail of carpentry offers the clearest evidence so far that the West, in some
measure at least, represents the archaic East.

Similar to V-notching, though lacking its diagnosticity, is the
saddle notch (Fig. 4.1). While many westerners use the term
"saddle" when describing V-notching, the two properly should
be distinguished as altogether different types of corner timber-

ing.[20] The saddle notch, used exclusively in round-log construction and always crowned, is formed by rounding out a saddlelike depression in the top or bottom, or both, into which the adjacent log fits (Figs. 4.16 and 4.17). Often the cut, if done by saw rather than axe, has slanted sides and a flat bottom, rather than a rounded shape, as is particularly common in New Mexico. The saddle notch, including this slant variety, appears throughout the mountain West, occasionally as the most common type, and accounts for another fifth of our field observations (Fig. 4.11).

Saddle-notching occurs in both the Midland Anglo-American and Mexican carpentry traditions, as well as among Scandinavians and Ukrainians. As a result it lacks diagnostic value. Still, something can be learned from its details. Undersided saddle-notching, less prone to catch and hold moisture, prevails in the Midland East, while the *oversided* subtype represents the most common western variety, some 56 percent in our sample.[21] The oversided type is easier to construct, and the axman usually fashions it while standing atop the structure being built.[22] We believe the cruder, oversided type represents the Midland frontier method, now rare and archaic in the East. If we are correct, then the prevalence of oversided saddle-notching in the West offers yet another example of archaic eastern influence. In some far northern districts, where the frontier imprint remains vivid, *all* examples of saddle-notching are oversided, as for example, the Mat-Su Valley and Arctic Circle–Steese Highway districts of Alaska.

Figure 4.15. Crowned V-notching on round logs, Silver City, Yukon Territory (research district 20). Such crowning is an archaic Midland Anglo-American carpentry feature common in the West but largely vanished from the East. (Photo by T.G.J., 1990.)

Figure 4.16. Oversided saddle-notching, Cashmere, Washington (research district 15). This represents an archaic eastern type in the Midland Anglo-American carpentry tradition. The structure is today in the Chelan County Pioneer Village. (Photo by J.T.K., 1991.)

Dovetailed Notching

Another fifth of all western corner timbering consists of well-crafted dovetailed notching, which occurs on both hewn and end-hewn logs (Fig. 4.18). Two distinct subtypes exist, *full* and *half dovetailing*. The former has two splayed surfaces on each log, the latter only one (Fig. 4.18). Both form superior, difficult-to-fashion, locking joints that delight the eye (Figs. 4.9 and 4.19). Half dovetailing is the more common type, and we found 222 examples, as compared with 146 specimens of full dovetailing.

Both dovetailed types occur in the Midland Anglo-American carpentry tradition, and one might be tempted to regard these notches as further evidence of Pennsylvania Extended. Closer analysis reveals a far more complicated cultural history. For one thing, full dovetailing is far less common in the Midland tradition than in the West and rarely diffused much west of Pennsylvania. It seems entirely likely that none of the full dovetailing in the montane West reflects Midland influence. Russians, Ukrainians, French Canadians, Scandinavians, and Finns all introduced this notch into the West, and fully one-third of all the full-dove-

Figure 4.17. Undersided saddle notching, Beaverhead County, Montana (research district 9). (Photo by J.T.K., 1989.)

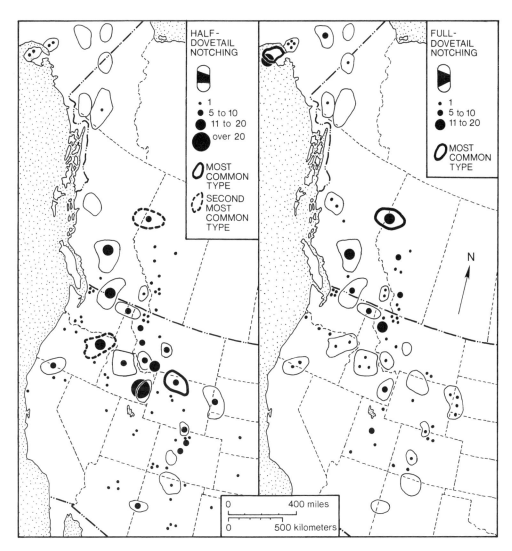

Figure 4.18. Distribution of dovetailed notch types.

tailed structures we found can be linked to those ethnic minorities.[23] In addition, full dovetailing also existed abundantly in the Anglo-Canadian carpentry of Ontario derived from the Yankee military tradition and was more common there than in Midland Anglo-America.[24] The remaining two-thirds of all full dovetailing in the West could well derive from the Anglo-Canadian East, and it is worth noting that over half of all nonethnic examples of this notch occur north of the 49th parallel. The landscape tells us, then, that Canada laid cultural claim to its own West, in substantial measure.

Similarly, the half-dovetail notch is very common in both Anglo-Midland and Anglo-Canadian carpentry. In fact, it accounts

Figure 4.19. Half dovetailing on end-hewn logs, near Dawson Creek in the Peace River country, British Columbia (research district 17). (Photo by T.G.J., 1990.)

Figure 4.20. Hispanic vertical double-notching, highland New Mexico (research district 2). This is the most common Hispanic method of corner timbering and derives, in this area, from Mexico. In Old Cienega Village Museum, Santa Fe. (Photo by T.G.J., 1994.)

for fully three-fourths of all log buildings across a wide swath of Ontario and seems to share a culture history with full dovetailing.[25] The probable eastern Canadian connection receives added strength from the fact that one-third of all western examples of half dovetailing occur in conjunction with end-only hewing, a trait highly suggestive of Ontarian influence. At the very least, western half dovetailing emphasizes the collective role of the Anglophone East in shaping the West. Only one example of half dovetailing in an ethnic context was found, involving a Russian-built structure at Fort Ross, California.[26]

Double and Single Notching

Only one other notch type accounts for as much as 10 percent of our western sample. While often treated as separate types, the *vertical double notch*, also called a "lock" joint, is best considered as one variety with subtypes. In about three-fourths of all examples, right-angle cuts are made in both the top and bottom of the log, leaving a crown intact (Fig. 4.20). In the variant form, usually called *single-notching*, the cuts are made only on the top or bottom, not both (Fig. 4.7). Forty-two percent of all double and single notching occurs in one research district—Hispanic New Mexico—and the type also appears abundantly in Mexico (Fig. 4.21).[27] Clearly, double notching often represents an ethnic type in the West, a fact reinforced by its occasional appearance in Russian, Finnish, and Scandinavian settlements.[28]

Still, the majority of western occurrences remains unexplained in ethnic terms (Fig. 4.22). In an earlier study, we concluded that the double notch often represents an archaic, crowned variety of Midland Anglo-American square notching, a type more common in frontier times and now largely vanished from the East.[29] The greatest concentration of Anglo-American double notching occurs in Alaska and the Yukon, the areas retaining the most pronounced frontier character. In two districts there, vertical double notching is the most common type (Fig. 4.21). In the far North, at least, double notching probably represents the archaic Midland East.

Minor Corner Types

A variety of minor cornering types exists in the West, most of which find their explanation in small ethnic minorities. A favored French-Canadian method, sometimes called Red River frame or *poteaux et pièces*, spread with the early fur trade and usually occurs at Hudson's Bay Company posts (Fig. 4.9). Instead of notching the timbers, French carpenters erect morticed, weight-bearing corner posts, into which the wall logs are tenoned. Additional, medial posting removes restrictions on the size of the building, permitting some impressive structures.[30] French-Canadian cor-

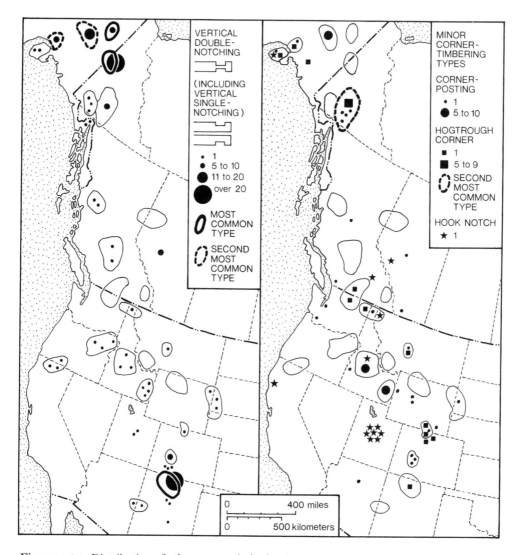

Figure 4.21. Distribution of other corner-timbering types.

ner and medial posting never spread appreciably beyond their group, and we found only thirty-seven examples in the entire West—less than 2 percent of the buildings (Fig. 4.21).

Even rarer is another unnotched technique called the *hog-trough* corner (Fig. 4.23). Milled boards on the order of 2-by-8s, nailed together into an L-shape, take the place of proper corner posts in this crude, unaesthetic type. Nails driven through the boards into the butt ends of the logs secure the corners. The hog-trough is the second most common cornering method in the upper Yukon Valley, and nearly half of all examples occur there, though a scattering can be found fairly widely (Fig. 4.21). The technique is apparently a western innovation.[31] Cruder still is

Figure 4.22. Nonethnic vertical double-notching, better revealed due to crown damage on one wall, Minto, Yukon Territory (research district 19). (Photo by J.T.K., 1990.)

Figure 4.23. Hog-trough cornering, Carcross, Yukon Territory (research district 19). (Photo by J.T.K., 1990.)

the *butt joint*, in which unnotched logs are spiked together (Fig. 4.24). Log barns in the Matanuska region of Alaska have butt cornering.[32]

Finally, a very rare cornering type, the *tooth* notch, appears in a few Finnish, Scandinavian, and Russian areas, most notably central Utah (Fig. 4.25). This difficult-to-fashion method, also called the *hook* or *tongue* notch, never diffused beyond the ethnic context (Fig. 4.21).[33] The most distinctive aspect of the tooth notch is a morticed projection on the bottom side of each log at the joint, formed with a saw.

Covering Log Walls

Regardless of which method of corner timbering they employed, western carpenters usually left the log walls exposed to the weather. Clearly, log construction did not bear the stigma in the West that existed in the postpioneer East. In this sense, too, the region remains loyal to its frontier roots, to its archaic legacy. A vigorous revival of unadorned log house construction has recently spread throughout the West, prompting a deluge of how-to manuals and journals.[34]

Exceptions to the dominance of uncovered walls are rare. False-fronts of milled lumber often conceal the logs of commer-

Figure 4.24. Spiked butt joint on log barn in Matanuska Valley at Palmer, Alaska (research district 24). The structure dates from the 1930s. (Photo by T.G.J., 1990.)

Figure 4.25. Tooth-notching in a Norwegian ethnic settlement at Sundre, Alberta, in the foothills of the Canadian Rockies. (Photo by T.G.J., 1990.)

Figure 4.26. Adobe plastering over hewn, double-notched log walls of a house in Hispanic highland New Mexico (research district 2). Unlike most Hispanic log dwellings, this one has chink construction. The structure stands in Old Cienega Village Museum, Santa Fe. (Photo by T.G.J., 1994.)

cial buildings from street view (Fig. 2.15). Only Hispanic New Mexicans consistently conceal the wooden walls of their dwellings, using an adobe plaster (Figs. 2.17, 3.28, and 4.26).[35] Mormons in Utah and Idaho occasionally apply an adobe covering, as do Ukrainians out on the Canadian prairies.[36]

As a result, grayed, weathered log walls remain a highly visual aspect of montane western landscapes, contributing appreciably to the regional character. One unfortunate result is that few western log buildings are not endangered. A frightful waste of a cultural resource progresses year by year in the high valleys of the West. We must not assume that the sheer abundance of log buildings in the western mountains assures a long-term future for these traditional structures. Even those log buildings now gathered in outdoor museums deteriorate unless maintenance is regularly provided.

Western Roof Structure

Several distinctive roof types occur in the West, each possessing some explanatory power. Nearly one-third of all roofs rest upon a kind of structure known as the *ridgepole-and-purlin* (Fig. 4.27).[37]

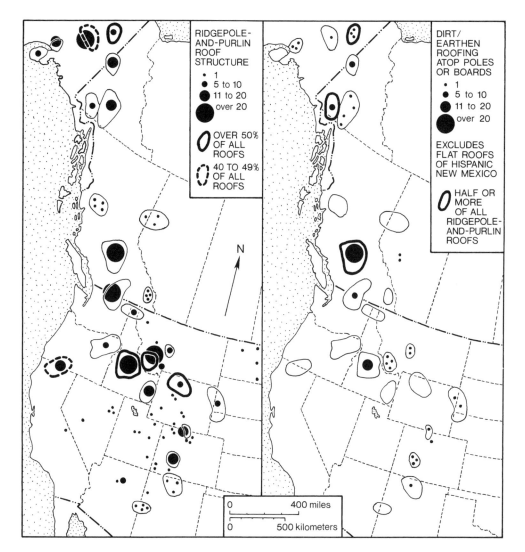

Figure 4.27. Distribution of ridgepole-and-purlin roofs and earthen roofing.

Achieved by carrying log construction up into the gable, it consists of a ridgepole and parallel purlins notched into progressively shorter gable logs. The roof supports, then, follow the axis of the eaves (Figs. 2.5, 2.7, 2.8, and 2.9). The result is a massive, very strong structure that can bear easily the weight of the roofing material and an insulating layer of winter snow.

Ridgepole-and-purlin roofs characterized the frontier stage of Midland Anglo-American carpentry and have since become rare east of the 100th meridian. Ease of construction, involving the same skills used in building walls, made the ridgepole-and-purlin an ideal frontier type. In the West, we can regard it as an archaic Midland form. It occurs throughout the far greater part of the montane West, occasionally as the dominant type (Fig. 4.27).[38]

Though the diffusionary link is clear, the ridgepole-and-purlin roof of the mountain West differs somewhat from its eastern prototype, mainly in having a lower pitch (Fig. 4.28). In fact, some "Anglo-Western" roofs rest merely upon a ridgepole and lack purlins altogether. In some other cases, the ridgepole is doubled for added strength, causing the roof to crest atop two closely spaced purlins that, in effect, take the place of a proper ridgepole (Fig. 2.19). Many other roofs have only one or two pairs of purlins to flank the ridgepole. As a result, the roof pitch can be 5 degrees or even less, though most fall in the 20-25 degree range. Roofs more closely resembling the eastern prototype, with up to five or six pairs of purlins can be found in the West, but on the average western roofs are more gently pitched than those of the East.[39] This may result from a desire to capture a thicker insulating snowcover.

Dominant in both the Midland and Anglo-Canadian East is a lighter, raftered roof, supported by poles or milled beams seated on the top, or "plate," logs in the eave wall and joined together at

Figure 4.28. Low-pitched ridgepole-and-purlin roof on a trapper's cabin, from Hot Springs County, Wyoming, and now relocated to Cody Old Town Museum (research district 6). The ridgepole rests on the same gable log as the purlins and stands higher than they do only by virtue of its greater thickness. This, then, is an "Anglo-Western" roof. (Photo by T.G.J., 1987.)

the ridge. Trusses add strength, as does lathing nailed at right angles to the rafters. Of Germanic origin, this raftered roof has a steeper pitch than the ridgepole and purlin, usually between 35 and 40 degrees, accommodating an attic (Figs. 2.4, 2.14, 2.20, 3.4, and 3.5). Gables consist of boards nailed to studs.

The raftered roof finds abundant representation in the West, appearing on just over half of all log buildings. They are especially common on barns, where attic hayloft space is important, and on second-generation houses.[40] Western mountain people use the term "attic and shingle roofs" to distinguish them from the older, purlin type.[41] A variant of the raftered roof, the *gambrel*, first appeared on barns in the East shortly after 1850 and went west abundantly (Fig. 3.8). Rather than consisting of two equal, unbroken pitches, the gambrel is trussed outward, forming a break in each pitch and enlarging the loft or attic.[42] A distinctive variant of the western raftered roof occurs in Hispanic New Mexico. Distinguished by an unusually steep pitch, these regularly exceed 45 degrees and can reach even 60 degrees (Fig. 3.24 and 4.29). Its origin remains uncertain and disputed, but most such roofs appeared after 1930.[43]

Still another variant of the raftered roof, the *hipped* or pyramidal type, appears widely if rather infrequently in the West (Fig. 4.30). Diverse eastern groups, including French-Canadians, Midwesterners, and Tidewater southern Anglo-Americans had known such roofs before coming to the West, as had Russians and Ukrainians.[44] The western pyramidal cannot, therefore, be diagnostic, and does not represent a western innovation.

Reflecting a highly distinctive ethnic tradition is the flat roof found on close to half of all Hispanic log buildings in highland New Mexico (Figs. 2.17, 3.18, 3.23, 3.27, and 3.28). Called an *azotea*, the Hispanic roof spans the structural unit with massive horizontal log *vigas*, or beams. Smaller members, called *latias*, rest atop the large beams and at right angles to them. A covering of small branches, twigs, grass, straw, sod, rock, or gravel is placed over the *latias* to form the watertight surface. Clearly in New Mexico the *azotea* represents the older, more traditional roof form, one with deep roots in both Spanish and Pueblo Indian culture. Even many Hispanic log structures with steep, raftered roofs prove on closer inspection to have a concealed *azotea* below. The attic in such cases usually has no function. In all likelihood, the *azotea* leaked too much during times of melting snow, leading the Hispanos to shelter it with a second roof.

The only other roof structure type occurring with any frequency in the West is the single-pitch, shed roof mentioned earlier (Figs. 3.13, 3.14, 3.15, and 3.16). About 5 percent of all western roofs are of the shed type. It almost certainly represents an archaic eastern type in Anglo-American areas.[45]

Figure 4.29. Steep-pitched board roof on a small Hispanic dwelling, highland New Mexico (research district 2). A flat roof is concealed beneath the gabled, raftered one. The house is in the Old Cienega Village Museum, Santa Fe. (Photo by T.G.J., 1994.)

Figure 4.30. Hipped, pyramidal roof on a two-room cabin, British Columbia, in the Vanderhoof Community Museum. Such roofs appear widely, if not very frequently, in the West. (Photo by T.G.J., 1990.)

Roofing Material

In the East, the typical covering for frontier ridgepole-and--purlin structures in the Midland Anglo-American subculture consisted of loose, unnailed shakes held down by weight poles. Spacers called knees kept the poles in place, and the lowest tier of knees rested against a cantilevered "butting log" atop each eave wall. Only three examples of this kind of roof have been documented for the entire West, none of which survives.[46] We must conclude that westerners adopted the frontier roof structure of the Midland East while rejecting its traditional covering. Second-generation, raftered log roofs in the East almost invariably had shingling, a type well represented in the western mountains today, but not the traditional covering (Figs. 2.4, 2.5, 2.18, 2.22, 3.5, and 3.6). Instead, the most common western roof, whether on ridgepole-and-purlin or raftered structures, once consisted of milled boards or, less often, board and batten (Figs. 3.3, 3.12, and 4.29). These planks, nailed to the purlins or lathing, run from ridge to eave. Sometimes rounded sawmill remainders, a variant of board-and-batten roofing, appear, and occasionally these are heaped in chaotic multiple layers (Figs. 4.31 and 4.32). Another favored traditional western roof consists of tight-spaced poles, often poplars, nailed to purlins (Fig. 2.11).[47]

Whether covered with boards, remainders, or poles, the western ridgepole-and-purlin roof often acquired an additional layer of soil, turf, or clay as much as a foot thick (Figs. 1.1, 2.11, 4.27, and 4.32). Sometimes a layer of cut grass or bark rested between

Figure 4.31. Board roof made of sawmill remainders, Robinson, Yukon Territory (research district 19). Most likely by coincidence, identical roofs occur in interior Finland. (Photo by T.G.J., 1990.)

Figure 4.32. Roof of stacked sawmill remainders, with dirt atop, on a derelict ridgepole-and-purlin structure near Gunnison, Colorado (research district 3). This roof covering represents an almost complete departure from eastern tradition. (Photo by T.G.J., 1987.)

Figure 4.33. Grass grows atop a dirt-covered ridgepole-and-purlin roof, near Cache Creek, British Columbia, in the largest regional concentration of such roofing (research district 16). Note also the end hewing and full-dovetail notching. (Photo by T.G.J., 1990.)

the wood and earth, and westerners refer to these as "pole, thatch, and dirt roofs." In time, grass usually sprouted on top, helping stabilize the roof (Fig. 4.33), but another solution was to place an additional layer of boards atop the dirt.[48] The largest concentration of dirt roofs occurs in central British Columbia (Fig. 4.27). Though high in maintenance needs, dirt provides excellent insulation.

Absent altogether from the East, earthen roofing would seem at first glance to be a western innovation. Some think it appeared first on the Great Plains, before settlers ever entered the mountain West.[49] Conceivably, Scandinavian settlers could have introduced the turf roof, since the type is well known there, especially in Norway. The earthen flat roof of Hispanic New Mexico, too, might have had some influence. More likely, we feel, the western dirt roof represents another borrowing from the Indian tribes of the region, some of whom lived in winter quarters covered in this manner.

Another distinctive western roof, found on stabling barns in one area of British Columbia, consists of haystacks atop a supporting platform of poles.[50] Such anomalies keep us honest. One can never be sure what is around the next bend of the road in the West, and the region exhibits more visual surprises than Turner, Webb, or any other armchair scholar ever dreamed.

The analysis of western log carpentry reinforces our tentative image of the West as a complex region of continuity, innovation, and diversity, a place at once archaic and inventive, mainstream and ethnic. Alongside carpentry features that demonstrably belong to Pennsylvania, Mexico, Ontario, Russia, or Scandinavia are other form elements developed locally or borrowed from Indians. Some carpentry features appear throughout the West; others occur in localized clusters. The West is obviously a complicated place culturally, and we are not yet ready to reach final conclusions. We must first seek out other message-bearing elements of the cultural landscape. Traditional wooden fencing, the subject of the following chapter, will enhance our understanding of the West.

Chapter Five

Wooden Fences

Traditional fences represent another building component of the western folk landscape. Robert Frost instructed us long ago that "good fences make good neighbors," but geographers know that fences can also serve as informants, helping us decipher and explain America's cultural patterns. They hold clues to diffusions, innovations, and origins. Fencing provides "a significant index of settlement stage and character" and is one of the few landscape features to "combine so finely the characteristics of the resource base, the cultural matrix and its historical antecedents."[1] Furthermore, since fence types exhibit pronounced regional variations, they can be as indicative of cultural patterns as carpentry and folk architecture.[2]

Many miles of traditional wooden fencing survive in the montane West, in spite of wind, flood, and fire. Both stockraisers and farmers helped fence the West; only the districts north of the agricultural frontier in Alaska, the Yukon, and British Columbia remained unenclosed (Fig. 1.4). Enclosure began before the advent of barbed wire in the mountains, with the result that various kinds of folk fences appeared on the landscape. True, barbed wire has pretty much taken over the region, but we were amazed by the amount of homemade fencing that survives. These "runs" of humble enclosures told us many revealing things about cultural origins and character.[3]

In addition, western wooden fences add a special character to the landscape, a picturesque aspect long exploited by Madison Avenue for "Marlboro Man" and other advertisements. The Fall, 1995, catalogue of The Territory Ahead, a Santa Barbara mail order firm, featured a mountain western wooden fence on its cover, and forty other regional fences of various descriptions graced the interior pages of the catalogue. Clearly, there is more to these structures than meets the casual eye.

Worm Fencing

As proved true of dwellings, outbuildings, and carpentry, certain montane western fence types have an eastern origin. In fact, the *worm*, or *snake*, fence, the most common traditional type in the forested East, is also most numerous among western wooden fences (Fig. 5.1). Consisting of round poles or split rails stacked in a zigzag course and normally lightly notched at the joints, this type derives from log construction. Originally introduced by Finns and Swedes in the colonial Delaware Valley, worm fencing spread by 1850 to become the national fence. It also became typical of the Anglo-Canadian East, in Ontario.

Several subtypes exist. The simplest and most archaic variety consists of round poles, held in place by gravity, the tripod principle, and shallow notching. In the postfrontier East, split rails replaced round poles, to the extent that the older type scarcely exists any longer in the cis-Mississippi lands. In addition, later generations of easterners added more lateral support to the worm fence, either by placing paired posts at each joint or by providing "stakes and riders" (Fig. 5.2). In the latter method, crossed diagonal stakes resting on the ground and forming an X-shape of braces were locked in place by an additional rail, or "rider," resting atop the crotch of the crossed stakes. All three of these subtypes occur in the West, but nearly two-thirds of all worm fences there represent the most archaic variety—round poles without supports.[4] The mountain West forms the last refuge of this pioneer subtype. In fact, the West houses the greatest survival of all types of worm fence. Another 17 percent of western worm fences involve unnotched round poles supported by a pair of posts at each joint, a subtype regionally called the "horse and rider" fence.[5] Its stronghold lies in the Salmon River country of Idaho.

Once common throughout most of the farming and ranching districts of the West, worm fencing today clings mainly to a few refuges in the interior Pacific Northwest (Fig. 5.3). Among our research districts, the principal survival focus lies in the Cariboo and Chilcotin plateau country of British Columbia, which yielded more than half of our field observations. Interestingly, in eastern North America, worm fencing also survives best on the Canadian side of the border, in Ontario.

Worm fencing in the West normally borders pastures or, occasionally, fields, but corrals are also built in this manner. The worm-fenced pens display an awkward star shape, achieved by making the angles of the zigzag alternately acute and obtuse.

Figure 5.1. Round-pole worm fence bordering a high pasture in the Uinta Mountains, northeastern Utah. (Photo by T.G.J., 1987.)

Figure 5.2. Posted worm pole fence made of aspen, also called a "horse and rider," Gunnison County, Colorado (research district 3). (Photo by T.G.J., 1987.)

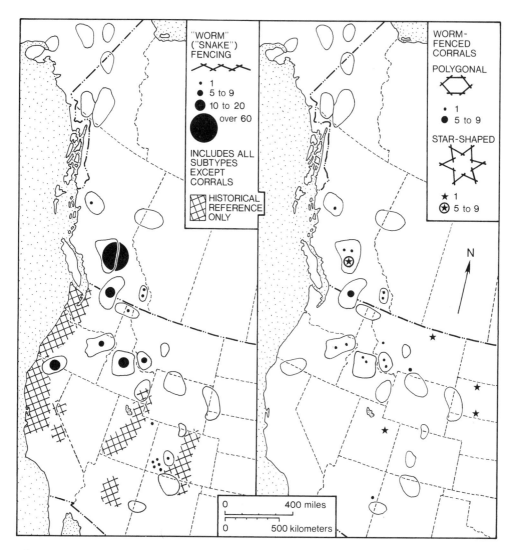

Figure 5.3. Distribution of "worm" fences and corrals in the montane West. The star-shaped corrals shown in the Dakotas, Montana, and Utah represent historic photographic evidence but they no longer exist.

The resultant enclosure, while roughly circular in configuration, has a succession of starlike points.[6] Few examples survive, in part because a more efficient western modification appeared (Fig. 5.3). It occurred to some pen builder in the northern or central Rockies that the corral fence need not zigzag, but instead could more efficiently shift inward at each joint, producing a simpler polygonal shape (Figs. 5.3 and 5.4). Fewer logs or poles and less labor were required, allowing the polygonal corral to eclipse its star-shaped ancestor. In earlier works, we mistakenly interpreted the polygonal rail corral as another archaic eastern type, since identical pens exist in Scandinavia, the European source region of worm fencing.[7] Our inability to find even one cis-Mississippi example or historical reference convinced us that the polygonal rail corral represents a western reinvention rather than an archaic eastern type.

The polygonal corral apparently originated in the Northwest, most likely in the Beaverhead country of Montana, an early center of ranching. Today, such pens are still being built along the northern perimeter of ranching in British Columbia, where cattle raising spreads into empty forest lands in one last episode of frontier expansion.

Other Eastern Fence Types

Closely related to the worm fence and also derived from notched log construction is another archaic eastern Midland Anglo-American type, the chock-and-log fence (Fig. 5.5).[8] Panels consisting of long logs alternate with short "chocks," allowing builders to erect a straighter fence than the worm type, while using fewer timbers. Logs rather than poles must be used, since

Figure 5.4. Polygonal corral employing the worm-fence principle, near Vernon, British Columbia (research district 15). This type apparently is a western modification of the worm fence to make a pen. (Photo by J.T.K., 1990.)

Figure 5.5. Chock-and-log fence, southwestern Montana (research district 9). Related to the worm fence, this type also has an eastern origin. (Photo by J.T.K., 1989.)

deep notching compensates for diminished tripod support. Though much rarer than worm fencing, the chock-and-log survives in about the same geographical refuge in the West (Fig. 5.6). Half of all extant runs of this fence type stand in British Columbia's Chilcotin-Cariboo district, a living museum of archaic enclosures.

Another traditional and largely vanished eastern Midland fence found in the West is locally called the *pitchpole, rip-gut,* or *stake-and-rider* (not to be confused with the worm fence subtype of the same name). The name "rip-gut" derives from the ominous-looking projecting poles at the top of the fence, which appear capable of disemboweling an animal. In common with the worm and chock-and-log types, the pitchpole represents the Midland Anglo-American subculture. Finns introduced the type from northern Europe to the Delaware Valley about 1650.[9] A "bristly" fence, the pitchpole consists of poles or rails laid at a slant, one end resting on the ground and the other in the crotch of X-shaped, crisscrossed stakes that are firmly fastened together to form a "buck." The structural principle of the pitchpole is mutual support, with the crossed stakes holding up one end of the rail or "rider," whose weight in turn holds down the bucks and prevents them from falling. In erecting a pitchpole fence, the work commences, in the words of a western builder, "by securing one end of a rail to a post or tree at the desired height," usually about five feet, "and letting the rail slope to the ground" at an angle of about 30 degrees. "Across this rail, about four feet from the upper end, are driven two stakes which form a crotch into which the next rail is laid," parallel to the first. "This simple procedure is repeated until the required length of fence is standing."[10]

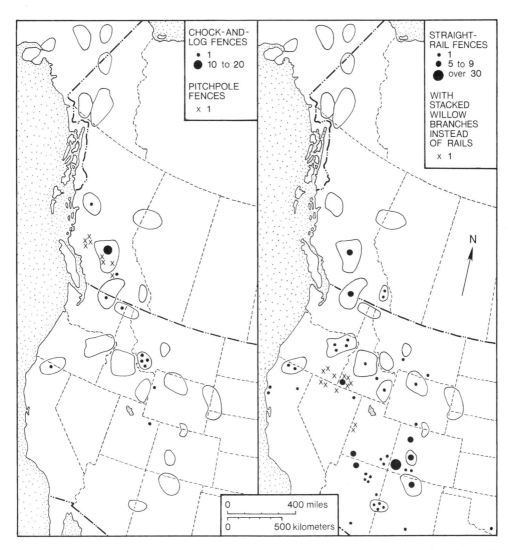

Figure 5.6. Distribution of chock-and-log, pitchpole, and straight-rail fences in the West.

The pitchpole goes by other, diverse names in the East, including "buck," "Swede," and "Irish" fence, but it has become extremely rare there.[11] Even out West, relatively few examples survive, though the pitchpole reportedly once appeared commonly in a belt from Montana to northern California.[12] We found pitchpoles only in southern interior British Columbia (Figs. 5.6 and 5.7). The type "seems to have achieved its greatest development in the Anahim Lake district of the far Chilcotin," where long stretches of it are to be seen, indicating "an unusual interest and confidence in the design."[13] Yet another archaic Midland fence, then, finds its greatest survival in the Chilcotin-Cariboo country.

Figure 5.7. Pitchpole fence in its last remaining western stronghold, the far Chilcotin country of interior British Columbia. An archaic eastern Anglo-American type with roots in colonial Pennsylvania, the pitchpole is now equally archaic in the West. (Photo by Donovan Clemson, from *Outback Adventures: Through Interior British Columbia* [Saanichton, British Columbia: Hancock House, 1976], p. 15; courtesy of Hancock House Publishers.)

The final example of an eastern Midland Anglo-American fence type to find a place in the West is the *straight-rail,* usually called a "post-and-rail" back East. It consists of pairs of posts set at about six- to eight-foot intervals, with poles or split rails stacked between the supports (Fig. 5.8).[14] Most often round aspen poles are employed in construction, presenting, as in the case of worm fencing, the more archaic eastern Midland subtype. Split-rail examples are only about half as common out West. Straight-rail fences usually form livestock pens, especially those used for cattle, but in places they also border meadows or pastures (Figs. 5.8 and 5.9). Stronger than worm fencing, this type could better withstand the shoving of cattle but was more labor-consuming to erect. Only modestly represented in the West, straight-rail fencing appears frequently only in southern intermontane regions, especially the Four Corners area and the Owyhee country where Idaho, Nevada, and Oregon adjoin (Fig. 5.6).

Figure 5.8. Straight-rail meadow fence, made of round poles, Eagle County, Colorado. (Photo by T.G.J., 1987.)

Figure 5.9. Straight-rail corral complex made of aspen poles, Gunnison County, Colorado (research district 3). This type often reflects a Texan influence by way of the cattle ranching industry. (Photo by T.G.J., 1987.)

While ancestrally linked to Midland Anglo-Americans in the East, straight-rail pens became most closely linked to the Anglo-Texan ranching culture and probably entered the West mainly with cowboys from the Lone Star state. The type was also known in frontier Missouri and in Mexico, presenting a complicated cultural history that denies the straight-rail any diagnostic potential.

A western innovation produced a distinctive variant of the straight-rail fence in one region, the High Desert of southwestern Oregon, southeastern Idaho, and northern Nevada (Fig. 5.6). In this subtype, willow branches and sticks are stacked between pairs of juniper posts, spaced about 8 to 14 inches apart and set at 3- or 4-foot intervals, producing a "willow corral."[15] So localized is this variant that not a single specimen occurred in our twenty-five research districts. Similar brush fences in Mexico cloud the issue of whether the willow corral results from a local modification or diffusion with Hispanic *pastores* and *vaqueros*.

The Buck: A Western Fence

Other traditional varieties of fence occur only in the mountain West. Distinct from all known eastern fences, these enclosure types instead appear to be of western origin. If not purely innovative forms, they at least represent very substantial modifications of eastern types. The first of these enclosure types is the western *buck* fence, sometimes also known as the "sawbuck" or "buck-and-pole" fence (Fig. 5.10). The fundamental component of this enclosure, the buck, consists of two 5- to 6-foot posts morticed or nailed together to form a single X-shaped unit much like the support of the pitchpole fence. Horizontal rails, the uppermost resting in the crotches of the bucks and the lower tiers nailed to one side of the bucks, stretch between each pair of bucks (Fig. 5.11). Builders frequently put an additional tier of rails on the back of the fence to increase its stability. This forms a sturdy but movable type of enclosure requiring no in-ground posts. Moreover, it consumes less timber than the worm fence because it does not zigzag across the landscape. Perhaps because of these advantages, the buck fence is the second most common type we found in the West, serving equally well as a meadow or pasture enclosure.

The western buck enclosure exhibits a fairly widespread distribution in the mountain West (Fig. 5.12). While locally intense concentrations occur in the Cariboo country of British Columbia and in the Upper North Platte River valley, the overwhelming focus of buck fencing appears in the Salmon River country of Idaho. There we observed no fewer than thirty-eight separate runs of buck fence, more than two and a half times as numerous as worm fences and representing the single most common type

Figure 5.10. Characteristic X-shaped *buck* provides the support for the fence of the same name, northern Custer County, Idaho (research district 10). (Photo by J.T.K., 1991.)

in this district. Linked to this focus was a secondary concentration across the Bitterroot Range in southwestern Montana, where buck fences also outnumbered worm enclosures. Western Wyoming appears to be another area in which this type is a very common landscape fixture.[16]

Confusion exists in the scholarly literature regarding the nomenclature of the buck fence. Some authors refer to it as the "jack" or "jack-and-pole."[17] Buck fence is the better term, if for no other reason, because western ranchers use that word.[18] It has existed as a type by this name for over a century. An account from the late 1800s in the ranch country of Wyoming's upper Medicine Bow Valley mentioned fences constructed of pine poles, "three or four inches in diameter and sixteen to eighteen feet long," affixed to "a buck formed by two pieces of timber eight to ten inches thick and six feet long, notched and fitted together at right angles."[19]

Clarity in nomenclature also sheds light on the origin of the buck fence and, by extension, of the regional culture of the West. The name "buck," coupled with the structure of the fence, strongly suggest derivation from the Midland pitchpole fence, which is known back East as the "buck," among other names. Both re-

Figure 5.11. Western buck fence bordering a cattle pasture, Lemhi County, Idaho (research district 10). (Photo by J.T.K., 1991.)

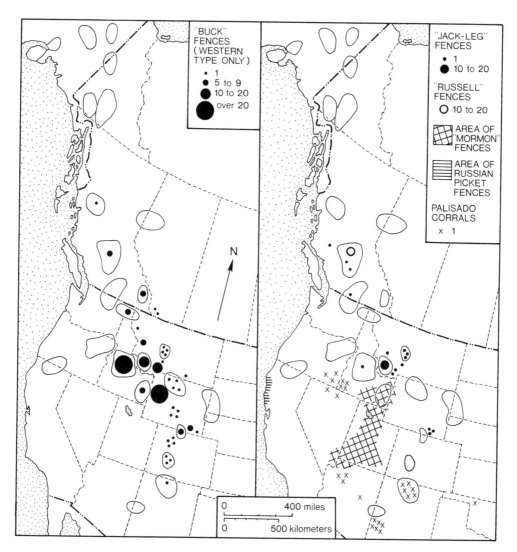

Figure 5.12. Distribution of buck, jack-leg, Russell, and various picket fence types. (The "Mormon" fence occurrence is after Francaviglia 1978b, pp. 29, 76.)

gional uses of the word "buck" refer to the similarity between the X-shaped supports and the design of the "buckstand," "buck-horse," "sawbuck," or simply "buck," the Midland American names for a sawhorse.[20] At the heart of both fences is the buck. A relatively minor western modification leveled the slanted rails of the pitchpole fence by affixing them directly on the bucks at both ends. This reduced the number of buck supports needed and took advantage of readily available, commercially manufactured nails.

The eastern stake-and-rider worm fence also had a buck and might conceivably have provided the inspiration for the western buck type, but it never reached the West in force, yielding to the more primitive unsupported worm fence. This freed the term "stake-and-rider" to be applied, in some western districts, to the similar pitchpole fence, and that, in turn, allowed the term "buck" to migrate to the new adaptation. These "musical chairs" name shifts and the rarity of the stake-and-rider worm fence in the West, when combined, point to the pitchpole as the direct ancestor of the western buck fence. Had it derived from the eastern stake-and-rider, the name likely would have remained with it.

Amid the confusion of names, we are safe in considering the western buck fence as a regional modification of an eastern Midland type. The distribution and chronology suggest that the innovation perhaps occurred in southwestern Montana, possibly as early as the 1850s among cattle ranchers of Midwestern origin. The place of origin may instead lie farther east, in the Great Plains of Kansas or Nebraska, in the western margins of the Midwest proper.

The Jack-Leg and Russell Fences

Very similar to the western buck fence, and often confused with it, is the jack-leg type. The two are clearly closely related. The jack-leg, like the buck, rests on two-legged supports, with a tier of horizontal poles attached with nails (Fig. 5.13).[21] The only difference is that the buck of the jack-leg consists of one thicker member, into which is morticed the tenon of a smaller pole. Western ranchers acknowledge the distinctiveness of this type by giving it the different name.

An early account from Madison County, Montana, described a fence built of posts which were not set into the ground; instead, according to the observer, "each post was stayed by a leg," hence the appellation.[22] The jack-leg fence looks a lot like a more refined fence type, requiring more substantial carpentry to construct. On some jack-leg fences, western builders even morticed and tenoned the horizontal rails into the larger posts.[23]

The distribution of the jack-leg, though similar to that of the buck, is more limited, and we found fewer examples of it in the mountain West.[24] The jack-leg fence is most prominent in the Beaverhead country of southwestern Montana, where examples appeared almost as often as buck fences and more abundantly than worm enclosures (Fig. 5.12).

The origins of the jack-leg fence, while undocumented, are almost certainly western. It perhaps evolved as a refinement of the buck fence, and its limited distribution is consistent with the confined spread typical of other western innovations. One piece of evidence contradicts this scenario. A traveler's account from the Great Plains in 1859 described a Kansas fence built by driving "stakes in the ground at an angle of about 60°," then inserting "from the opposite side another stake through it by means of

Figure 5.13. A jack-leg fence beside an irrigation ditch in the Big
Hole Valley, southwestern Montana (research district 9). (Photo by
J.T.K., 1989.)

an augar hole and nailing on the strips which rest on the
stakes."[25] The jack-leg may have been invented there on the
prairie, as a wood-saving response to a largely treeless environment, in which case the western buck fence becomes an easier-
to-construct degeneration of the jack-leg, occurring after about
1870. In any case, we must not overlook the Great Plains as a
source area for western innovations and adaptations.

Similar to the buck and jack-leg fences is another related, distinctly western type, the *Russell* fence. Also supported by X-
shaped bucks and formed of stakes wired together, the Russell
variant, too, has one upper pole resting in the crotches of the
supports (Fig. 5.14). Its distinctiveness comes from the lower
rails being suspended in wire stirrups hanging from each buck.[26]
"Swiftly and easily erected," the Russell fence often soon sags
and looks derelict, but cattle tend to respect rails better than
wire and will usually stay clear.[27] Despite the often flimsy appearance, "many miles of Russell fence in use in the ranching
country today are proof of the soundness of the design and its ef-
fectiveness as a barrier to stock."[28]

Overall, the Russell fence is not a major western type. We observed fewer than twenty examples in our sample areas. Like
other western innovations, the Russell fence has singular regional focus (Fig. 5.12), and all of the specimens we found lay in
the Cariboo and Chilcotin country, the most traditional ranching

district in British Columbia. Donovan Clemson earlier noted this concentration.[29] The "Russell fence straggles across much of the Cariboo country. It is the hardy perennial that gives to the country a special character."[30] He also concluded that the Russell fence has been a part of the Cariboo and Chilcotin cultural landscape since just after the turn of the twentieth century, often appearing in old photographs.

Clemson, a fence builder himself, attributes the development of the Russell fence to western ingenuity.[31] It probably evolved as a local subtype of the buck fence in the Cariboo and Chilcotin country, when some individual found it easier to suspend the horizontal rails from wire loops than to nail them to the faces of the bucks. Certainly the highly localized distribution of the Russell fence favors such an origin. Clemson further suggested that the Russell fence may have emerged as an effort by western ranchers to repair dilapidated worm fences without acquiring additional poles or rails. Some of the many miles of Russell fence in British Columbia are "no doubt" converted worm fences, "considered beyond repair." These "supplied enough material for a new Russell fence to take its place." The Russell type "requires 5

Figure 5.14. A Russell fence in the Chilcotin ranching area west of Williams Lake, British Columbia (research district 16). (Photo by J.T.K., 1991.)

rails to the panel," while "8, 9, or more" are needed for worm fencing. Moreover, "the Russell fence is built on the straight," while the worm type "zig-zags, so the rails cover less distance." These advantages ensure the Russell's "continued popularity."[32]

All of the western fences employing the X-shaped supports, including the buck, jack-leg, Russell, and pitchpole types, share an adaptive advantage. They minimize the problem of rotting that plagues the fencing of wet meadows. Worm, chock-and-log, and straight-rail fences all proved vulnerable to this problem of standing water and decay. The X-shaped buck exposes the minimal amount of wood to rot. Moreover, the buck, jack-leg, and Russell fences greatly reduce the number of buck supports needed in the pitchpole type, with the result that the prototypical eastern form gave way to better adapted western modifications.

Picket Fence Types

Relatively few other traditional fence types appeared in our research districts. We encountered an occasional cluster of distinctive types, most commonly lying between the areas of intensive research. For example, an Hispanic-derived *palisado* corral type, made of closely-spaced juniper or scrub pine poles set vertically as pickets, appears in the High Desert of southeastern Oregon, northern Nevada, and southwestern Idaho, in pretty much the same area where willow branch corrals occur (Fig. 5.12).[33] Also called "picket" or "stockade" corrals, these pens typically have stout posts every 7 to 10 feet, with smaller stakes between; two horizontal strands of wire woven in and out provide additional strength. Small board strips sometimes take the place of the wire. Most likely, Hispanic buckaroos from California introduced this traditional Mexican fence into the High Desert. It also appears in the Hispanic New Mexico highlands area, and *pastores* from that area could be responsible (Fig. 5.15).

A similar picket fence, apparently of very different ethnic origin, occurs along the northern California coast, from Fort Ross north through Mendocino County (Figs. 5.12 and 5.16). Supported by posts positioned about every 6 feet, this fence contains a horizontal board to which are nailed closely spaced vertical members. Given their similarity to fences found throughout most of Slavic eastern Europe, we can tentatively ascribe this Californian type to Russian influence, derived from the early nineteenth-century Fort Ross colony.

More widespread is the distinctive picket enclosure known as the *Mormon fence* (Fig. 5.12). Builders affix horizontal upper and lower rails to cedar or juniper posts, then attach a disorderly array of scrap lumber as the pickets. "The emphasis is upon variety," and the result presents "an unpainted potpourri of different types and shapes of wood." In one fence, "as many as five differ-

Figure 5.15. Hispanic corral-barn complex with both picket and straight-rail fences at Lower Colonias, San Miguel County, New Mexico (research district 2). (Photo by C.F.G., 1971.)

Figure 5.16. Board or slat picket fence, near Elk, Mendocino County, California. This distinctive local type may reflect a Russian colonial influence. (Photo by T.G.J., 1992.)

ent picket styles" may appear, occasionally supplemented by a broken wagon wheel or hay rake.[34]

We have by no means exhausted the diagnostic potential of traditional western fences. Gates, for example, offer an interesting topic for future research (Fig. 5.17). But, incomplete though it may be, our fencing analysis has again demonstrated the diversity and paradox of western culture: at once ultraconservative, even archaic, in its preservation of certain traditional eastern features, and at the same time devoted to change, adaptation, and innovation. It is almost time for us to try to make sense of this apparent confusion, but first we need to consider a final material cultural complex—that associated with haying.

Figure 5.17. Innovative gate design on a jack-leg fence, Beaverhead County, Montana (research district 9), attests to western inventiveness. (Photo by T.G.J., 1987.)

Chapter Six

MATERIAL CULTURE OF HAYMAKING

As a settlement frontier, the mountain West confronted farmers and herders with environmental conditions unlike those in the eastern United States and Canada. Overall, the mountain West was colder and drier than the eastern areas from which most Anglo-American settlers to the region came. Such environmental conditions appear to have encouraged modification of certain adaptive strategies, including the abandonment of the corn-farming and swine-raising agricultural system of the Midland backwoods folk.[1] New strategies arose to meet local environmental conditions. When faced with unfamiliar conditions in which known techniques or strategies produced less than satisfactory results, some western settlers responded creatively by modifying an existing item or method or by coming up with something innovative.

Nowhere is western adaptation and innovation more strikingly revealed than in items related to the production and storage of cut hay, large amounts of which were needed to feed herds through the winters. Hay as a feed crop had been known in much of the eastern United States, but primarily as supplemental feed. In the West, this crop became more important to stock ranchers because the long winters of the mountains limited grazing by sheep and cattle on natural forage to a smaller part of the year.[2] In fact, settlers soon realized that the more hay they could produce, the better. Isolated in their mountain valleys, western ranchers could do little but hope for chinook winds and watch their animals starve if they ran out of hay before spring. "Always, the ultimate dread of ranchers who fed hay was that the hay supply would be exhausted before grass was ready for grazing in the spring."[3]

Their response was to employ irrigation, greatly increasing meadow yield. Innovations were required in harvesting and storing hay, evidence of which remains on the landscape today. "In the East, farmers generally placed their hay in the barn loft," but in the mountain West there was just too much hay for such a strategy.[4] Furthermore, the basic repertoire of eastern material culture was inadequate to deal effectively with these large hay crops. Accordingly, western stockraisers began developing new storage techniques and machines to assist them in this task, producing "the mechanical revolution of the western range."[5] This inventory of hay-related material culture represents the major innovative western component of the built landscape, consisting of implements for stacking, enclosing, and allocating loose hay.[6] Different adaptive responses occurred in localized settings, then competed with each other for wider regional acceptance, and we can usually recognize western innovations precisely by their occurrence in individual regions, whereas items introduced from the East reveal wider distributions.[7]

Hay Stackers

The method of storage that became the standard, at least into the early 1900s, was to stack the hay loosely in the open air. Hay stacking was not, in itself, an innovative technique, as it had been practiced in Europe and back East. Stacking, however, had rarely if ever been employed with such large amounts of hay. Fortunately, the same semiarid conditions that prompted the use of irrigation aided western ranchers in their hay stacking. If constructed properly, these stacks could keep for four or five years, giving ranchers a lot of flexibility in their storage and allocation. This method was not foolproof, as even well-stacked hay experienced a certain degree of spoilage, especially on the top of the stack. To minimize the loss, ranchers had to pile it as high as possible, because the taller the stack, the less the spoilage. To create such stacks, some device was necessary to elevate the hay to the top, once it became higher than workers could reach. To accomplish that, western ranchers invented hay stackers.

Diverse types of hay stackers appeared in the mountain West, and "it was probably in the types of stacking devices that ranchers displayed the greatest amount of ingenuity."[8] Eventually, only two forms dominated. The first of these utilized an inclined plane—a ramp or slide—to elevate hay to the top of the stack, while the second principal form employed a derrick with a pivoting boom or arm mounted on a stationary base of some kind, usually a post or a tripod (Figs. 6.1 and 6.2). Both types developed gradually as the products of evolutionary processes, acquir-

Figure 6.1. Slide stacker of the Beaverslide type, near Wisdom in Montana's Big Hole Valley (research district 9). The device was invented in this very valley. (Photo by J.T.K., 1991.)

Figure 6.2. Derrick-style hay stacker, at Frenchglen in the High Desert of southeastern Oregon. (Photo by T.G.J., 1991.)

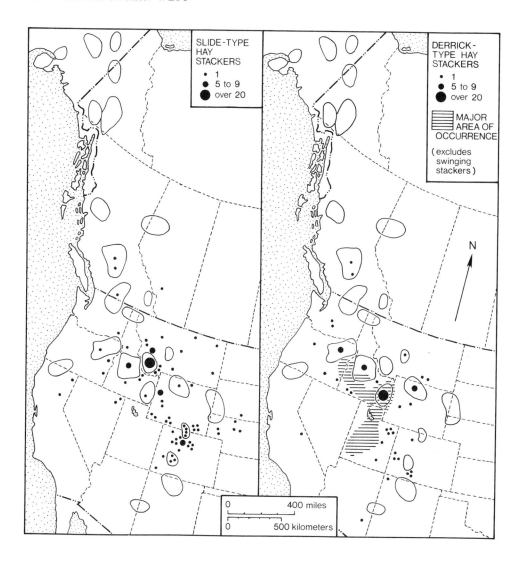

Figure 6.3. Distribution of slide and derrick hay stackers. (In part after Alwin 1982, p. 47, and Francaviglia 1978a, p. 921.)

ing a number of subtypes or variations. They also had distinctive geographical distributions within the mountain western region, as one would expect of innovations.

The slide or ramp stacker eventually achieved a widespread distribution in the West (Fig. 6.3). It apparently began as a device consisting of two 45- to 50-foot-long poles, placed about 15 or 20 feet apart and roughly vertical at the edge of a developing haystack. The poles angled toward each other slightly, so that their upper ends stood only about 6 or 7 feet apart, connected by a horizontal crosspiece. Workers slung a large net between the poles, which they lowered to fill with hay. The hay crew raised the load with ropes and horse power, tilted the derrick from the pivot of its base until it hung directly over the top of the stack, and then dumped the hay from the nets onto the stack by means of a trip cord. H. S. Armitage, a resident of the Big Hole Valley in the Beaverhead country of southwestern Montana, described the use of such a device as early as 1891.[9]

This crude derrick proved cumbersome and unwieldy, and western ranchers began tinkering with it. Perhaps even earlier, certain other Great Basin ranchers had built "simple wooden inclines on which net loads of hay were pulled up," so that "the stack was built back from the ramp."[10] An old illustration of this primitive type of inclined plane stacker in the Blue Mountains of Oregon exists, but we found none surviving on the landscape.[11]

Then, in the Big Hole Valley during the 1890s, someone combined the twin-pole derrick concept with the idea of the inclined plane. In this new device, called the "ram" stacker, the load of hay was raised up the slide by a large, horse-driven plunger, consisting of a long pole with a plowlike blade for pushing the hay (Fig. 6.4).[12] Armitage described the use of such a slide stacker in the Big Hole Valley by about 1903. "The top end was 16′ or 18′ above the ground. The hay was pushed over the top by a long plunger pole, with the horses working on the back end, as far from the stack as the slide was long; then the team turned around to pull the plunger down off the stacker, ready for the next load."[13] The efficiency of this "plunger" or "ram" slide stacker prompted its modest spread to other parts of the mountain West, especially through the central Rockies (Fig. 6.3). It reached the

Figure 6.4. Slide stacker of the plunger, or ram, type, near the North Platte River, Wyoming (research district 4). (Photo by J.T.K., 1993.)

upper North Platte Valley of Wyoming by the 1920s.[14] Other examples of this prototypical slide stacker have subsequently been
observed elsewhere in southern Wyoming, as well as in Colorado.[15]

The ram or plunger stacker formed an intermediate stage in
the stacker innovation process. While more efficient than the
two-pole derrick, it had disadvantages, not the least of which
was the height of the haystacks produced. The top end of the
ram stacker rarely stood taller than twenty feet, meaning the
haystacks themselves could be no higher. Since a larger ramp,
and thus a longer plunger, was impractical, the next best option
was to raise the height of the stacker by increasing the angle of
the slide. This apparently made use of the plunger mechanism
difficult and unstable. To solve these problems, ranchers in the
Big Hole Valley devised the more refined Beaverslide stacker,
also known as the Beaverhead County or Sunny Slope Sliding
Stacker (Figs. 6.1 and 6.5).[16]

The Beaverslide preserved the inclined ramp design with its
twin 50-foot-long poles. Supporting these timbers, and making
for a much steeper slide, were taller wooden supports, between
20 and 30 feet tall. This increased the angle of the ramp to nearly
45 degrees. To raise the hay up the slide, the inventors added a
large basket consisting of wooden teeth to hold each load. Crews
raise this basket by means of horse power applied to a system of
cables and pulleys, until it reaches the top of the slide and dumps
the hay onto the stack. Because of this innovation, the awkward
plunger with its long pole became unnecessary.

In addition to its height advantage, the Beaverslide is much
more mobile than the ram stacker, since it forms a self-contained
unit without a separate plunger. In fact, western ranchers usually
place their Beaverslides on skids so they can move them from
meadow to meadow to produce the numerous haystacks characteristic of the high valley western landscape. Stacks of more than
twenty tons of hay nearing 30 feet in height are commonplace.
These advantages, combined with the speed with which crews
could stack hay, made the Beaverslide a highly successful piece of
agricultural technology.[17]

As a result, the Beaverslide experienced a rapid and fairly
widespread diffusion through the West (Fig. 6.3). From its single
point of origin, the Big Hole Valley, where two ranchers produced the prototype in 1908, the Beaverslide spread through adjacent parts of Montana and Idaho, along the Rocky Mountain
chain into Wyoming and Colorado, and as far afield as northern
California, British Columbia, and the Nebraska Sand Hills. Even
so, Beaverslide stackers remain most numerous in their place of
origin, the Beaverhead country, where we observed twenty-nine
examples. A secondary concentration occurs just to the west

Figure 6.5. Beaverslide hay stacker in the Big Hole Valley of southwestern Montana (research district 9). (Photo by J.T.K., 1991.)

across the Continental Divide, in Idaho's Salmon River country, where a half-dozen Beaverslides stand in the Lemhi Valley south of the town of Salmon. The Beaverslide may have achieved its impressive western diffusion in part due to the efforts of seasonal laborers in the haying industry, who moved from one ranching area to another and spread the concept of the stacker. Big Hole residents themselves often traveled throughout the West constructing Beaverslide stackers, and one individual reputedly built five hundred Beaverslides.[18]

Over the years westerners continued to refine the Beaverslide stacker, making it still more efficient. They added fences along the sides of the ramp to prevent hay from falling off as the basket was raised. Ranchers also devised free-standing backstops to help hold and shape the haystack. Later, some ranchers also added shorter backstops to the sides of new haystacks, helping to stabilize them still further.[19] Some of the more modern refinements include the use of gasoline-powered engines to raise the hay basket and the construction of metal versions of the stacker (Fig. 6.6). Many remain in use today, and we observed functioning Beaverslides in the Big Hole Valley, in Lemhi County, Idaho, and in Teton County, Wyoming.

That the Beaverslide represents the culmination of an ongoing process of innovation in haying material culture, and specifically a more advanced version of the "ram" stacker, seems certain. The visual similarities and working principles of the two stackers are virtually identical. More important, the ram stacker was known to the inventors of the later stacker by the name Beaverslide.[20] The unique value of the Beaverslide stacker to our analysis of mountain western culture lies in the fact that this invention is well documented, both in terms of date and place. We may assume that all western innovations followed this model.

Derrick Stackers

Derricks with moving arms, or booms, represent the second basic type of western hay stacker, also the product of local innovation (Fig. 6.7). They lift the load of hay and swing it out over the stack. Like the slide stacker, the hay derrick experienced an evolutionary development following an initial western invention, producing a number of subtypes.[21]

Hay derricks emerged from a simple original apparatus—a stationary vertical pole supported by guy wires. Cables and pulleys allowed the crew to drag loads of hay up onto the stack. Subsequently, but still before 1900, westerners added a horizontal mast and an angled boom to this design, producing some of the earliest hay derrick subtypes. These modifications allowed cleaner lifting of the hay loads and more precise placement on the stacks. Further modification produced still more refined derrick types, with much longer booms either mounted on top of shorter, sturdier masts or slung under tripodlike supporting frameworks (Fig. 6.8). Besides producing sizable stacks, these subtypes were also more mobile. The sturdier booms of the later subtypes allowed larger, more quickly constructed hay stacks, while their greater mobility meant shorter hay transport distances to the stacking sites. In this respect, the derrick stacker experienced a formulative development virtually parallel to that of the slide stacker.[22] The hay derrick, then, provides another example of western innovation in the process of creating a modified adaptive strategy of land use. It became "one of the many ingenious 19th-century folk agricultural innovations" that help give a special character to the West.[23]

The geographical distribution of the hay derrick strengthens the evidence that the device is a western invention (Fig. 6.3). Rather than occurring throughout the region, the hay derrick is heavily concentrated in the eastern Great Basin. The Fifes found some fifteen hundred of them in that area in the 1940s, a concentration later confirmed by Richard Francaviglia.[24] Residents in much of the western United States routinely refer to the hay derrick as the "Mormon derrick" or the "Mormon stacker." Francaviglia concluded on the basis of its overall distribution, evolutionary development, and terminology that the western hay derrick was a Mormon invention, developed and spread by that group as they settled the Great Basin in the late 1800s. Referring specifically to the distribution of derrick subtypes, he felt there

Figure 6.6. Beaverslide stacker in use near Jackson, Teton County, northwestern Wyoming. A gasoline-powered engine has replaced horsepower for lifting the hay up the slide. Note the back and side-stops used to shape the haystack. (Photo by J.T.K., 1991.)

Figure 6.7. Western hay derrick east of Savona, British Columbia (research district 16). Note the swivel post and boom design. (Photo by J.T.K., 1991.)

Figure 6.8. Underslung hay derrick northeast of Weiser, Washington County, Idaho, on U.S. Highway 95. (Photo by J.T.K., 1991.)

was "a hay derrick hearth in the farming valleys of central Utah," the core of Mormon Deseret, where the most primitive forms still appear.[25] More complex and refined forms, of later origin, occur with greater frequency in the Mormon peripheries.[26] This conclusion is compatible with our own findings in the field. Granted, our research did not take us into the heart of the Mormon country in central Utah, but nevertheless most of the hay derricks we observed lay in the northern fringe of the Mormon culture region, mainly in Idaho. Most represented later subtypes, in keeping with Francaviglia's findings. Outside of the Mormon culture region, we found only scattered examples of western hay derricks.

The slide stacker, especially the Beaverslide, prevails in the Rocky Mountain region while the derrick represents the dominant type in the intermountain West, between the Rockies on the east and the Sierra Nevada/Cascades to the west. As embodiments of one of the West's most important innovations—the storage of cut hay in massive outdoor stacks—these two rival indigenous designs served to compartmentalize the innovative aspects of western culture into subregions. They provide landscape evidence of two parallel haying systems, each suited to a different environmental niche. The first system, based on the harvest of native hay, was one in which the speed of harvesting was most important.[27] The slide stacker, an innovation ideally suited to mountain valleys where wild, native hay grew in relative abundance, became the stacker of this system. The Beaverslide is an inexpensive stacking implement which can quickly stack large amounts of hay, an important virtue in areas with extensive and often rough wild hay meadows.[28] It served large ranching operations in which the quantity of stacked hay was more important than quality. The Beaverslide, because of the manner in which it dragged hay up the ramp, did not always preserve the quality of the cut grass, but ranchers did not see this as a problem. In fact, the Beaverslide was most commonly used in conjunction with the buck rake, a horse-drawn device used to gather the dried windrows of hay and transport them to the stacking site. The rake, notorious for producing a lower-quality fodder, rolled and tumbled the hay.[29] Because of its speed, the buck rake, like the Beaverslide, made up in quantity what it cost in quality. Both machines proved ideally adapted to gathering the wild hay of Rocky Mountain valley bottomlands.

The Mormon hay derrick of the Great Basin, on the other hand, served a much different agrarian system and ecological niche, one in which alfalfa hay replaced wild species and the quality of the feed was paramount. In the drier intermountain West, native hay did not grow in nearly the abundance typical of the mountains. Since winter fodder was equally necessary here, farmers seeded meadows with alfalfa and irrigated. Requiring a more intensive operation and yielding a higher-quality forage, these alfalfa crops could not undergo the same treatment as wild hay during transportation and stacking. The buck rake, for example, often knocked the leaves off alfalfa. In response, the Mormon settlers of the intermountain West developed the derrick. It allowed for a much cleaner and gentler lifting of the hay onto the stack. In fact, where ranchers converted their meadows from wild hay to alfalfa, they often changed from using slide stackers to hay derricks.[30] Similarly, alfalfa hay crews employed a different, gentler method to transport hay from the windrows to the stacking site. Instead of buck rakes, workers utilized hayslips—flat, horse-drawn skids onto which cured hay was pitched by hand. Once towed to the stacking site, hay was removed from the slips with the derricks and set directly on the stack.

While commercially made hay stackers later became available, often employing fundamentally different mechanisms and designs, the splendidly adapted slide and derrick folk designs never gave way in the mountain West. The catapult *overshot* stacker, a mass-produced type, swept the Great Plains but never became dominant in the Rockies and Great Basin.[31] Similarly, the *swinging* stacker, a commercially made derrick device able to pivot laterally at its fulcrum and equipped with a hay basket at the end of the boom, never seriously challenged the homemade models. In their 1940s traverse of Utah and eastern Idaho, the Fifes observed only two commercial models among some fifteen hundred hay stackers.[32] During our fieldwork, we observed only two overshot stackers and four swinging stackers, all in Mormon Idaho. As a result of a questionnaire sent to Great Basin ranchers, Young learned that most respondents felt manufactured stackers were not strong enough for western hay conditions.[33]

Hay Cribs, Sheds, and Racks

The haystacks standing in the meadows of the mountain valleys and interior plateaus of the West must be protected from livestock during the fall. Ranchers enclose some of the larger ones with buck and jack-leg fences, but another means of protecting stacks is the roofless hay crib (Fig. 6.9). This small, open-chinked structure of loosely-fit, notched logs allows adequate ventilation while preventing livestock from depleting their winter feed prematurely. Another functional aspect of these cribs is the ease with which they can be dismantled, moved, and reassembled at a new haystack location (Fig. 6.10).

Very few roofless log hay cribs survive on the western cultural landscape. Wyoming's North Platte River valley has the only noteworthy concentration (Fig. 6.11). The modern distribution suggests a western invention that achieved only a modest diffusion, but such an impression is probably false. Instead, these small cribs were likely far more widespread formerly and represent an archaic eastern item of material culture. Through much of the southern Appalachians, small haystacks were traditionally built around a vertical pole, which provided stability, and enclosed by worm or straight-rail fences.[34] In form, these closely

Figure 6.9. Roofless, notched-log hay crib, near Encampment, Carbon County, Wyoming (research district 4). (Photo by J.T.K., 1989.)

Figure 6.10. Recently relocated empty roofless hay crib, Carbon County, Wyoming (research district 4). The month is July, and the cutting of hay lies ahead. (Photo by T.G.J., 1987.)

resemble the montane western roofless crib. The decline of the log hay enclosure likely began when larger haystacks could be produced by slide or derrick stackers. Indeed, it is remarkable that any at all survive, given their unsuitability to large-scale western haymaking. They represent an archaic eastern feature now also archaic in the West.

Even in decline, the roofless log hay crib underwent modification in the West. To afford the haystack more protection, some ranchers erected a steep, open-gable roof over the structure, in effect converting it into a haybarn (Fig. 6.12). Even in this modified form, the hay crib never became common.

A far more successful eastern implantment in the mountain West is the stilted hayshed (Fig. 6.13). Essentially a roof supported by tall posts, with open sides, this shed finds its prototype back East, where identical structures were built in Yankee states such as Michigan at least as early as the 1830s.[35] The more remote progenitor was likely the hay *barrack*, introduced by Dutch colonists into the early Hudson Valley and later passed to their Yankee New Englander neighbors.[36] The barrack's roof, also supported by corner poles, could be elevated as the haystack grew, a feature absent in its crude offspring, the stilted hayshed.

Older versions of the stilted hayshed found in the mountains

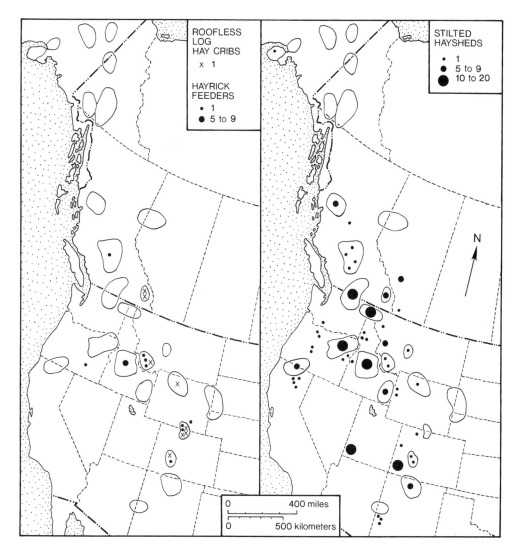

Figure 6.11. Distribution of roofless log hay cribs, log feed racks, and stilted hay sheds.

of the West tend to be small, but the structure could easily be built in much larger dimensions, allowing it to survive the western revolution in hay production. Even very large stacks could be sheltered in such sheds, reducing hay loss on the top of the stack. Many stilted sheds have a board skirting at the bottom to prevent livestock from gaining access (Fig. 6.14). In its enlarged form, the stilted hayshed has spread in recent decades throughout much of the United States and southern Canada, but its traditional stronghold remains the mountain West. The lack of a clear geographical concentration in any part of the montane region further suggests an eastern origin (Fig. 6.11).

Figure 6.12. Log hay crib covered by a steep-pitched roof, south of Quesnel, British Columbia (research district 16). (Photo by T.G.J., 1990.)

Figure 6.13. Stilted hayshed, a traditional design that survives in modern construction, in Mormon eastern Idaho (research district 7). (Photo by T.G.J., 1987.)

Figure 6.14. Stilted hayshed with skirting at the bottom to prevent pilferage by livestock. Note also the shed enlargement. It stands in Okanogan County, Washington (research district 15). (Photo by T.G.J., 1990.)

Figure 6.15. Log feed rack in Montana's Beaverhead country (research district 9). Hay is placed in the funnel to minimize waste. (Photo by J.T.K., 1989.)

Ranchers also sought to minimize waste in the distribution of hay to the livestock. They "treated hay in the stack like money in the bank," feeding as a last resort and as little as possible.[37] To prevent too much hay from being trampled or covered by snow-fall, ranchers devised feed racks, locally called *hayricks*.[38] These small, elongated log cribs above which are mounted funnel-shap-ed bins made of slats or poles, allow hay to be distributed with minimal waste (Fig. 6.15).

The geographical pattern of hayrick occurrence suggests a western innovation, probably in southwestern Montana or adja-cent Idaho (Fig. 6.11). Design and construction details vary con-siderably from place to place. While normally of log construc-tion—undoubtedly the prototypical form—hayricks sometimes lack the notched base, in which case the funnel bins reach to ground level.

The repertoire of hay-related material culture in the montane West, in common with folk architecture, carpentry, and fencing, speaks clearly to the issue of the origins of western culture. Quite simply, this repertoire represents the most innovative component of the region's built environment. With few excep-tions, the hay-related material culture of the West was developed by settlers within the region. Most often, these innovative items represent responses to new environmental conditions. Specifi-cally, the severity of the winters in the mountain and intermoun-tain West required ranchers to put up sufficient winter feed or risk losing their herds of sheep and cattle to starvation. Because of the sheer quantity of hay this involved, ranchers needed to de-velop new devices to assist them. The result was a collection of items uniquely western.

The time has now come to attempt to bring together the var-ious messages obtained in our reading of the cultural landscape of the western mountains. What, cumulatively, do these artifacts tell us about the sectional culture?

Chapter Seven # THE WEST REVEALED

What, in the final analysis, is the significance, if any, of these diverse and often quaint wooden relics? Are we three merely romantic, ivory-tower antiquarians who escape to bucolic, archaic places in an effort to avoid the increasingly unpleasant and often absurd real world? Are we better grouped with collectors of Shaker furniture, Appalachian quilts, or southwestern kachina dolls than with serious scholars of the West?

Well, we freely admit to being fortunate romantic escapists, regretting only having been born into an age far later than we would have chosen. Yes, we are antiquarians monkishly hiding in the happy grove of academe. No, our labors will not cure cancer or secure world peace. But we are no more escapists than those who hide in archives and libraries. We find legitimization in our first-hand knowledge of the land we seek to explain. These artifacts, which we have studied so lovingly, allowed us to glimpse truths about the West never before articulated or demonstrated.

While debate on the character and genesis of North American western culture must continue, the message of the built landscape seems clear enough, if disquietingly complicated. Traditional wooden structures tell us that the West is at once indigenous and imported, innovative and ultraconservative, Anglo-American and ethnic, unitary and plural. A final assessment of these diverse, contradictory, and paradoxical traits is now in order.

Continuity

The West cannot be understood without attention to cultural diffusion from the Anglo-American East. Pennsylvania Extended, borne by the Midland backwoods frontier culture complex, finds abundant representation in the mountain West. In part, as

Frederick Jackson Turner proposed more than a century ago, the Midland eastern influence appears in archaic forms. Diagnostic items of Midland Anglo-American culture wholly or largely vanished from states east of the Mississippi appear abundantly in the West. Examples include the "Anglo-Western" gable-entry cabin, the ridgepole-and-purlin roof, crowned V-notching, predominantly round-log construction, the pitchpole fence, chock-and-log fencing, the round-pole worm fence without stakes and riders, the roofless log hay crib, and the straight-rail pole corral.

The greatest single refuge of archaic Midland traits is British Columbia's Cariboo-Chilcotin country, followed by the Salmon River, southern Cascades, Okanogan, Beaverhead country, and middle Tanana districts (Fig. 7.1). The preeminent position of the Cariboo-Chilcotin country in Canada as a refuge of archaic, Midland-derived items calls to mind the fact that Ontario preserves these same elements far better than, say, Ohio. Something rooted in the Anglo-Canadian culture must explain this repeated international discrepancy in level of historic preservation. The far North, too, is a major reservoir of archaic features. Considering that these boreal districts in the Yukon Territory and Alaska lie north of the agricultural frontier and, therefore, possess no fencing or hay-related material culture, the number of surviving archaic Midland features found there becomes extraordinary.

Other, nonarchaic items of Midland Anglo-American origin also proved abundant in the montane West, as perhaps best exemplified by uncrowned V-notching and the front-gable single-crib barn. These nonarchaic features display a wider geographical range than do relict forms, achieving in some cases an almost ubiquitous presence in the West that pays no heed to the international border (Fig. 7.1). Cumulatively, the Midland Anglo-American imprint on the West is profound and pervasive. Pennsylvania Extended is a potent force to be reckoned with in seeking a cultural understanding of the West.

Be that as it may, the western landscape offers another clear message concerning cultural diffusion from the East, a clue to causation never previously noticed. Certain items of traditional western material culture came not from the Pennsylvanian hearth, but instead from the composite Anglo-Canadian–Yankee New England culture complex that had become rooted in upstate New York and Ontario in the early nineteenth century. Drawing upon form elements of the colonial military garrison architectural heritage of early New England as well as from the Pennsylvanian culture complex, the Anglo-Canadian–Yankee group developed a distinctive hybrid log building tradition back East. Examples of this tradition that successfully went west include end-only hewing, a devotion to dovetailed forms of corner notching, particularly the full dovetail, and the stilted hayshed.

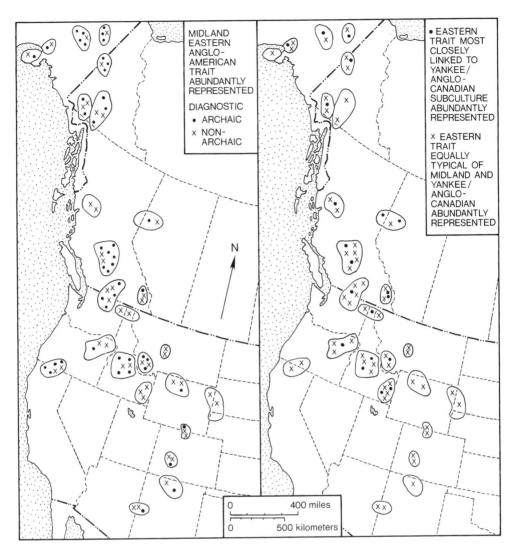

Figure 7.1. Mapping eastern cultural continuity in the West. Only data for the twenty-five research districts are shown.

While most common on the Canadian side of the border out West, these Anglo-Canadian–Yankee features spill south into the United States (Fig. 7.1). We should remember that the bearers of this subculture showed little respect for the international boundary in their migrations. As a result, British Columbia resembles Idaho, just as, a century earlier, Ontario had much in common with New York. The message of the western landscape is clear. Anglo-western culture cannot be explained solely in Pennsylvanian terms. Ontario, too, is relevant. If the West is culturally diverse, part of the reason is that the Anglo-American East possesses a parallel, equally profound, and kindred plurality.

Some eastern influences on the western landscape defy categorization as Pennsylvanian or Ontarian, including form elements common to both of these eastern Anglo subcultures. Half-dovetail and square notching belong in this group, as do shed-roofed cribs, the English plan cabin, chinking, and round-log construction. All of these also illustrate continuity from the East and display a far-flung western distribution, but they lack more specific cultural-historical diagnosticity (Fig. 7.1).

The list of items of eastern Anglo-American material culture that achieved successful diffusion to the mountain West, then, is impressive. Equally striking, however, is the inventory of eastern items that failed to achieve a noteworthy presence in the West. Examples include "dogtrot" and "saddlebag" house types, as well as transverse-crib and bank barns. Because these major eastern types are absent or very poorly represented, western folk architecture takes on a rather different aspect. We should pay as much attention to the failure to diffuse as to successful transplantation. At the root of both is cultural preadaptation.[1] Items of material culture that survived the westward migration were normally those that functioned as well or better in the environment of the West as they had in the East. Overall, carpentry techniques— log construction and the various shaping, fitting, and notching techniques—most easily made the transition from East to West. They were preadapted for success in the new land. Architectural style and design, whether the plan of a house or barn, often fared poorly. Styles declined or were abandoned quickly if ecologically inappropriate in the western setting, as, for example, eave-entry houses and barns, which had disadvantages in snowy climates.

So, lest we celebrate eastern frontier continuity too much and be tempted to become Turnerians, remember what did *not* come West. Know that a pragmatic sifting and winnowing of eastern culture occurred along the road west. A lot of cultural baggage was jettisoned along the trail.

Ethnic and Indigenous Continuity

Continuity in the cultural formation of the West entails more than Anglo-American and Anglo-Canadian influences. Pennsylvania and Ontario were not the only hearths, and the roles played by other Euroamericans must also be considered. Hispanos in highland New Mexico successfully implanted many elements of log construction from old Mexico, including distinctive notch types, chinkless fitting, adobe plastering, and horizontal gristmills with flat roofs. Among the highland Hispanos, too, carpentry most easily achieved diffusion. We need to recognize that the region we Anglo-Americans culturocentrically call the "West" was *el Norte*, "the North," to Hispanics. For the Russians, it was the far East, and they also left an ethnic mark on its cultural land-scape, as did the Finns, Scandinavians, and French Canadians.

As a general rule, such ethnic influences remained confined to the groups that introduced them. Anglos borrowed little from other Euroamericans. The result, geographically, is a veritable polynesia of small ethnic enclaves in the West, each challenging our broader assumptions about the region and demanding par-ticularistic explanations. Euroamerican continuity in the West was messy and geographically complicated (Fig. 7.2).

Another startling message delivered by the western cultural landscape, and one rarely noticed previously, is that while Ang-los did not often borrow culturally from their fellow Euroameri-cans, they most assuredly did so from American Indians. The stilted food cache and earthen roof tell us as much. Western cul-ture owes part of its distinctiveness to this interplay between the native peoples and the largest group of settlers. Too often we lament the passing of native groups and cultures, failing to real-ize that they helped shape the succeeding way of life. In the West, if we believe the landscape evidence, the native American contribution was greatest in the far North (Fig. 7.2). At the same time, native groups in the West typically adopted elements of Euroamerican material culture and often displayed them in quite distinctive ways, as in the notched-log, polygonal Navajo *hogan.*

Innovation

The conservatism or even ultraconservatism of the West, re-flected in its diverse continuities, finds a heavy counterbalance in innovation and adaptation. Westerners tinkered, invented, modi-fied, and diversified, in an ongoing process surely as old as the Euroamerican presence in the region. Not one adaptive strategy brought into the West worked flawlessly in the new habitat. Change had to occur. Only part of that change could be achieved by resurrecting rarely used items in the existing repertoire of material culture, such as the gable-entrance cabin. Most adapta-tion required genuine inventiveness, and the Webbian West be-came a far more profound hearth of invention than the East had ever been. For early colonial Europeans arriving on America's Atlantic shore, adaptation involved mainly a winnowing of arti-facts and practices derived from diverse European subcultures, a process of simplification and selection rather than innovation. These colonial Europeans had crossed an ocean only to find a new land startlingly like the one they had abandoned. In the West, by contrast, an altogether new habitat demanded ingenu-ity instead of simplification.

The cultural landscape clearly reflects western innovations and modifications. The mountain shotgun cabin, several types of

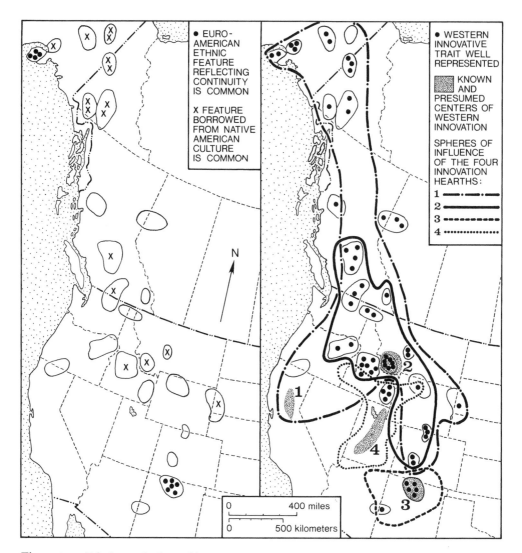

Figure 7.2. Ethnic continuity and innovation in the mountain West.
Only data for the twenty-five research districts are shown. The mon-
tane innovation hearths are: 1. California gold fields; 2. southwestern
Montana high valley ranching; 3. Hispanic highland New Mexico; 4.
Mormon Deseret. We omit the Great Plains but recognize that various
innovations may have occurred there, such as the jack-leg fence.

large gable-entry barns, nonconformist house plans, haystackers,
log feed racks, hog-trough cornering, oversized single-pen houses, and fences of the buck, jack-leg, and Russell types all bear
witness to the process. Cumulatively, these new forms help lend
a special character to western landscapes, while at the same time
revealing one of the basic cultural formative processes at work in
the region. Western innovation was disproportionately concentrated in the realm of livelihood, in the winning of life's necessi-

ties from the land. We would expect as much, since adaptive strategies deal with survival. The making of hay, fencing of land, and husbandry of livestock account for the large majority of western inventions revealed in the wooden structures on the landscape.

Western inventiveness displays a striking geography. Almost every new or substantially modified element of material culture remained confined to a particular part of the West, failing to achieve regionwide acceptance. The example of haystackers is instructive. In part, this geographical patterning occurred because different rival inventions arose to meet the same western need, but it also happened because innovations served particular subregional requirements.

Inventiveness occurred widely throughout the West. No single hearth of western culture arose. While we cannot pinpoint the precise source of most western landscape innovations, we believe evidence points to four montane centers or hearths, two of which coincide with our field research districts (Fig. 7.2). *Southwestern Montana*, the earliest center of high-valley cattle ranching, served as the principal hearth area of inventions serving that livelihood.[2] While the adjacent Salmon River country of Idaho has as high a rate of adoption of innovations as southwestern Montana's Beaverhead country, the much longer occupation history of the latter area, dating to the 1850s, and its acknowledged role in the development of western ranching point to the Montana area as the hearth. And the only western invention we can pinpoint in origin—the Beaverslide haystacker—came from southwestern Montana.

Mormonism may, as Ralph Waldo Emerson said, have been an echo of Yankee Puritanism, but the *Deseret* stronghold of the Latter Day Saints proved to be a significant center of innovation in material culture. Terms such as "Mormon fence," "Mormon cowshed," and "Mormon hay derrick" suggest their inventiveness.[3] Most innovations by the Deseret colonists occurred mainly in Utah and never spread beyond the confines of the Mormon culture region. Our research district in Mormon Idaho does not, therefore, constitute part of this hearth of innovation, but instead forms a periphery that borrowed from the Utah core area, much as the Salmon River did from the Beaverhead country. Deseret became an ethnic enclave, heightening internal regional contrasts in the West.

Hispanic New Mexico also functioned as an ethnic innovation hearth, remaining even more encapsulated than Deseret and yielding the most profound example of localized cultural distinctiveness in the mountain West. Bringing from Mexico an array of log carpentry techniques and architectural floorplans, the Hispanos developed from these an unruly array of new house and

outbuilding types that, as we saw, resists classification or even co-
herent nomenclature. While Hispanic carpentry methods spread
to some neighboring Pueblo and Navajo Indians, the highland
Spanish tradition borrowed and gave little.

The *California gold fields* hearth area, centered in the Sierra
Nevada foothills on the eastern side of the Sacramento Valley, is
both important and problematic. Its cultural landscape today retains surprisingly few genuine artifacts of the early mining era,
so few in fact that we could not justify it as a field research area.
What meet the eye there today are the much-gentrified artifacts
of the later, hard-rock commercial mining period. Photographic
evidence of gold rush times, however, leaves little doubt that California provided the prototype for the wooden built environment
associated with the mining frontier throughout the mountain
West. Even districts as far afield as the Yukon Territory bear the
mark of California's pioneer mining landscape.

We exclude the famous Willamette Valley of Oregon as a center of innovation, even though it was home to one of the earliest
introductions of Anglo-American culture in the West. Like the
California gold rush country, the Willamette Valley retains very
little traditional material culture and is, for that reason, difficult
to evaluate as a hearth of western landscape innovation. Its well
known role as a source of eastward-moving pioneers who settled
sizable parts of the interior Northwest suggests that it might
have been the scene of some western adaptations, but we find no
landscape evidence and no geographical distributions of material
innovations that point back to the Willamette. The Midland Anglo-Americans who settled there found it so familiar an environment that no substantial alterations of their adaptive strategy
were required. The Willamette Valley is, after all, a temperate,
fertile oasis on the margins of the hardscrabble West. That is
why settlers crossed half a continent to reach it, bypassing the
mountain West in the process. We should not expect the Willamette to have been a hearth of innovation, but instead a bastion
of continuity and eastern Midland conformity. The Willamette
Valley offers, in our view, a classic example of documented history contradicting the evidence of the cultural landscape. We believe the landscape.

Multiple Wests

The separate western hearths of innovation, the multiple eastern
and ethnic imported cultures, and the divergent western forms of
livelihood combined to produce, in the cultural geographical
sense, multiple Wests.[4] These several regions, vividly etched
into the cultural landscape, warn against belief in a Webbian,
monolithic West. Not even the montane part of the West forms a
single cultural region, but instead a mosaic of subregions. No

two among our research districts exhibit the same array of material culture, and only rarely would we feel comfortable in lumping several adjacent districts together as a larger region. Indeed, some districts, such as the Kenai Peninsula, need to be subdivided, separating, in that example, a Russo-Aleut western half from an Anglo-American eastern part. Some districts are devoted to farming, others to ranching, and still others to nonagricultural pursuits. Some became centers of innovation, while others served as refuges for archaic eastern material culture. Some exhibit a strong ethnic tinge, but most are mainstream American. Some bear a stronger Anglo-Canadian imprint, while others belong clearly to Pennsylvania Extended. Spheres of influence of the four innovation hearths often overlap confusingly. Expect, then, for each valley, basin, and plateau to house a distinctive subculture. Draw your own map of western subregions—we are not so foolish as to try.

In closing, we reiterate our underlying conviction: what is true of the traditional wooden structures in the built landscape is true of the culture at large. Our old buildings mirror the West, and the image they reveal is of a bewildering assemblage of unique places, born of continuity, innovation, and ethnicity. Both Turner and Webb were right about the West; both were also horribly mistaken. As for us, as soon as the snows recede, we will be back in the field, looking for "neat old stuff," undoubtedly learning that the mountain West is even more complicated culturally than we had imagined.

Appendix: Museums and Archives

Alaskaland Pioneer Park, Fairbanks, Alaska. Very large collection of log buildings from Fairbanks area.

Alaska State Library, Juneau, Alaska. Photograph collection.

Anchorage Museum of Art and History, Anchorage, Alaska. St. Michael's Russian log octagonal blockhouse, built 1833, on loan from the University of Alaska, Fairbanks.

Big Horn County Historical Society Museum, Hardin, Mont. Log cabin relocated to site.

Big Salmon Telegraph Office, Carmacks, Yukon. Cabin relocated to this site.

Calgary Heritage Park, Calgary, Alberta. Large collection of Albertan log structures, relocated to site.

California State Library, Sacramento, Calif. Photograph collection.

Willis Carey Historical Museum, Cashmere, Wash. Log structure.

Chelan County Historical Museum, Pioneer Village, Cashmere, Wash. Log structures.

Chugach National Forest, Kenai Peninsula area, Cultural Resource Files, on file at Anchorage, Alaska. Data on log structures inventoried.

Circle District Museum, Central, Alaska. Log structures.

Cloudcroft Museum, Cloudcroft, N.Mex. Four log structures.

Cody Old Town Museum (also called "Old Trail Town"), Cody, Wyo. Sizable collection of log buildings from Big Horn Basin.

Cooper Landing National Historical District, Cooper Landing, Alaska. Multiple in situ cabins, post office.

Dalby Pioneer Heritage Museum, Dalby, Mont. Pioneer dogtrot cabin.

Dalton Post Historic Site, Dalton Post, Yukon. In situ ruins of log building complex.

Dawes County Historical Society Museum, Chadron, Nebr. Log cabin and barn.

De Bolt and District Pioneer Museum, De Bolt, Alberta. Log cabin and
barn from local area.

Denver Public Library, Western Collection, Denver. Photograph collection.

Dunvegan Peace River Post Historic Site, Dunvegan, Alberta. Roman
Catholic mission and Hudson's Bay Co. log structures, in situ.

Eklutna Village Historical Park, Eklutna, Alaska. Log chapel exhibiting
Russian influence.

Ferry County Museum, Republic, Wash. Log structure.

Fort Crook Museum, Fall River Mills, Calif. One log house.

Fort Kenay Museum, Kenai, Alaska. Small collection of local log structures.

Fort Missoula Museum, Missoula, Mont. Collection of log structures
from local area.

Fort Ross Historic Park, Fort Ross, Calif. In situ Russian log buildings.

Fort Steele Heritage Park, Fort Steele, British Columbia. Large collection of log buildings from East Kootenay Trench.

Fort Walla Walla Museum, Pioneer Village, Walla Walla, Wash. Seven
log structures plus a replica of a blockhouse.

Museum of the Fur Trade, Chadron, Nebr. Two log dugouts, one other
structure.

Gakona Lodge and Trading Post, Gakona, Alaska. Log roadhouse and
diverse outbuildings, in situ and largely functioning.

Grand County Museum, Hot Sulphur Springs, Colo. Five log structures, including court house, school, and jail.

Grand Encampment Museum, Encampment, Wyo. Log structures from
Upper Platte River country.

Grande Prairie Pioneer Museum, Grande Prairie, Alberta. Five log
structures, including houses, church, and post office.

Gunnison Pioneer Museum, Gunnison, Colo. 1878 log cabin.

High Desert Museum, Bend, Oreg. Facsimile of 1869 cabin.

Idaho Historical Museum Pioneer Village, Julia Davis Park, Boise, Idaho. Two log houses.

Idaho State Historical Society, Archives, Historic Sites Survey, Boise,
Idaho. Photograph collection.

Klamath County Museum, Klamath Falls, Oreg. 1864 log cabin.

Klamath National Forest, Oak Knoll Area, Yreka, Calif. Site records of
log structures.

Log Cabin Heritage Museum, Kremmling, Grand Co., Colo. Log horse
barn, log jail.

Jack London Centre, Dawson City, Yukon. One log cabin.

Long Valley Museum, Roseberry, Idaho. Three Finnish ethnic log
cabins.

MacBride Museum, Whitehorse, Yukon. One cabin and a telegraph
office.

Charlie McConnell Ranch Site, Robinson, Yukon. In situ, derelict complex of seven diverse log dwellings and outbuildings.

Matanuska Colony Village Museum, Palmer, Alaska. Matanuska log
barn.

Matthews House Museum, Biggs, Carbon Co., Wyo. Double-pen log house.

Montague Roadhouse, Montague House, Yukon. Ca. 1900, derelict, in situ, with storage shanty behind.

Museum of New Mexico, Santa Fe, N.Mex. Photo collection.

O'Keefe Historic Ranch, Vernon, British Columbia. Log ranch house and horse barn, in situ.

Old Cienega Village Museum (El Rancho de las Golondrinas), south of Santa Fe, N.Mex. Large, nicely restored complex of log buildings from Hispanic highlands of New Mexico.

Old Log Church Museum, Whitehorse, Yukon. Log church and rectory, 1900.

Old Town Museum, Burlington, Colo. Log buildings.

Olmstead Place State Park Heritage Area, Ellensburg, Wash. Two log houses from 1870s.

108-Mile House Heritage Site, near 100 Mile House, British Columbia. Diverse in situ log buildings, including houses, hotel, barn, shed, ice-house, and a huge facsimile horse barn.

Oregon Historical Society, Portland, Oreg. Photograph collection.

Owyhee County Historical Complex, Murphy, Idaho. Beautifully preserved Anglo-Western log cabin.

Dorothy G. Page Museum, Frontier Village, Wasilla, Alaska. A small collection of log cabins, barn, and sauna.

Father Pandosy Mission, Kelowna, British Columbia. Eight log structures, including dwellings, smithy, chapel, barns, some in situ.

Pend Oreille County Historical Museum, Newport, Wash. Log buildings.

Perkins House Museum, Colfax, Wash. 1870 log house.

Pioneer Park, Ferndale, Wash. A collection of particularly well crafted log houses, plus a church and store.

Pony Express Station Museum, Gothenburg, Nebr. Log building, ca. 1860.

Rika's Roadhouse and Landing, Big Delta State Historical Park, Delta Junction, Alaska. Five in situ log barns, houses, roadhouse.

Rocky Mountain House Museum, Rocky Mountain House, Alberta. Anglo-Western log cabin.

Roop's Fort, Memorial Park, Susanville, Lassen Co., Calif. 1854 fort house.

Russian Bishop's Palace, Sitka, Alaska. In situ Russian log structure, beautifully crafted.

Millard Saylor Cabin, Ellensburg, Wash. Cowboy's Anglo-Western log cabin.

Robert Service Cabin Museum, Dawson City, Yukon. One log cabin.

Silver City Historical Site, Silver City, Yukon. In situ, derelict, numerous log buildings.

Siskiyou County Museum, Yreka, Calif. Two log houses, one ca. 1856.

Slate Creek Historical Museum, near Grangeville, Idaho. 1909 ranger cabin.

Soldotna Pioneer Museum, Soldotna, Alaska. Two cabins and a food cache.

South Dakota Historic Inventory, 1988. Data on log structures, provided by Carolyn Torma, Historical Survey Coordinator, University of South Dakota, Vermillion, S.Dak.

Stevens County Historical Museum, Colville, Wash. Log structure.

Stringham Cabin, a preservation exhibit in Ashley National Forest, Daggett Co., Utah. In situ Mormon log cabin.

Sundre Pioneer Museum, Sundre, Alberta. Norwegian log house.

Talkeetna Historical Society Museum, Talkeetna, Alaska. One trapper's cabin.

Terrace Heritage Park, Terrace, British Columbia. Nice collection of log structures from the Skeena-Coastal Mountains district.

Three Forks Pioneer Museum, near Albion, Wash. 1883 log house.

Vanderhoof Community Museum, Vanderhoof, British Columbia. Pyramid-roofed log cabin.

Van Slyke Museum, Caldwell, Idaho. Two log cabins, ca. 1864, from Canyon Co., Idaho.

White Mountain Historical Park, Springerville, Ariz. Three log houses.

Walter Wright Pioneer Village, Dawson Creek, British Columbia. Log house, school, and barn.

Notes

Chapter One The North American West: Continuity or Innovation?

1. Turner 1990, p. 12; Young 1970, p. 155; Meinig 1972, p. 159; Gressley 1986; Wyckoff and Dilsaver 1995, pp. 1–5.

2. Webb 1931.

3. Pomeroy 1955, p. 579.

4. Bloom 1972, pp. 157–59.

5. Meinig 1965, p. 197.

6. Turner 1893, p. 208.

7. Simpich 1923, p. 421; Pomeroy 1955, pp. 579–82.

8. Worster 1985, 1987, 1992; Limerick 1987; Limerick et al. 1991. For a good discussion of the "new" western history, see Wyckoff and Dilsaver 1995, pp. 15–20.

9. August 1993; Jordan 1992; White 1991; Wunder 1994.

10. Attebery 1976, p. 36.

11. Wyckoff and Dilsaver 1995, pp. 1, 7; Meinig 1972.

12. Kniffen 1965; Meitzen 1895, vol. 1, p. 28; Sauer 1930, p. 275. For more information on Meitzen, see Jordan, Domosh, and Rowntree 1994, p. 123.

13. Meinig 1979, p. 12.

14. Brumbaugh 1933, p. 6.

15. Kniffen 1960, p. 21; Poulsen 1982.

16. For preliminary statements that grew into this book, see Kilpinen 1991; Jordan 1991; Gritzner 1971.

17. Pomeroy 1955, p. 595.

18. Freed 1992, pp. 10, 153, 254.

Chapter Two Log Dwellings

1. Meitzen 1882, p. 3.

2. Kniffen 1965; Candee 1976; Alspach 1994; Stokes 1963; Rempel 1980; Jordan 1985.

3. Jordan 1985, pp. 24, 28, 31; Jordan 1994b; Jordan and Kaups 1987.

4. Wilson 1984, pp. 64, 66; Pitman 1973a, pp. 122–23, 154; Miller 1974, p. 113; Stedman 1985, pp. 34–35; Jordan and Kaups 1989, pp. 186, 189; Reeve 1988, p. 55.

5. Fife 1988, p. 135.

6. Davis 1989.

7. Jordan 1985, pp. 23–26.

8. Attebery 1982; Pitman 1973a, pp. 138–46; Jackson and Lee 1978, pp. 112–14; Gray 1989, pp. 110, 135; Dallas 1985, p. 98. Note that Okanogan is the Washington spelling, while Okanagan is used in British Columbia.

9. Kilpinen 1995a, pp. 20–22; Wilson 1984, pp. 12–14; Gritzner 1971, p. 61; Jordan and Kaups 1989, pp. 196–208; Davidson 1977; Reed and Smith 1982, p. 138; J. Attebery 1985, pp. 74, 158, 208; Kilpinen 1990, p. 100.

10. Wilson 1984, pp. 33–44; Kilpinen 1995a, p. 29; Reed and Smith 1982, pp. 45, 67.

11. Reynolds and Doll 1990, pp. 52–53; Hudson 1975, p. 10; Nelson 1986, p. 165; Freeman 1958, follows p. 64; Kariel 1992, pp. 148, 150; Davidson 1977, pp. 4, 11; Krenzlin 1965, pl. 4.

12. Wilson 1984, pp. 63–67; Welsch 1980, p. 319; Kilpinen 1995a, p. 28.

13. Jordan and Kaups 1989, pp. 199, 203, 206–8; Jordan 1985, pp. 24–27, 30, 32–33.

14. Jordan 1991, p. 14; Matheson 1985, p. 70; Evans 1988, p. 17.

15. Gray 1989, p. 152; Dallas 1981, p. 15; Miller 1974, p. 125; Engbeck 1990, p. 85; Roops Fort museum (see appendix).

16. Sultz 1969, p. 33; Miller 1974, p. 100; Chappel 1990, pp. 123–32; Gray 1989, p. 137; Weis 1988, pp. 155, 207; Yukon Historical and Museums Association 1983, pp. 25, 28; Schar and Pearl 1973, cover illus.; Bunting 1976, p. 110; Jones 1978, p. 9.

17. Wilson and Kammer 1989, pp. 20–24, 32; Reeve 1988, p. 24.

18. See examples in the Old Cienega Village Museum, Santa Fe, N.Mex. (see appendix).

19. Mattila 1971, p. 14; Walls 1989, p. 1; Lehr 1980.

20. Jett and Spencer 1981; Walls 1989, p. 1; Attebery 1991, p. 18.

21. MacBride Museum, Whitehorse, Yukon Territory, photo collection, vol. 47, no. 3966 (see appendix).

22. Reece 1985, pp. 14–15; Miller 1974, p. 150.

23. Russell 1979, p. 249.

24. Boyd 1958, pp. 220, 223; Gritzner 1971, p. 60.

Chapter Three Log Outbuildings

1. Jackson 1952, p. 31.

2. Francaviglia 1972, p. 153; for examples of studies on eastern barns, see Ensminger 1992, Arthur and Witney 1972, Wacker 1973, Jordan 1985, and Noble 1984.

3. Jordan 1985, pp. 30, 32–33.

4. Sandoval 1986, pp. 42–43.

5. Jordan 1985, pp. 30–34; Alspach 1994, pp. 94–106; Arthur and Witney 1972, pp. 60–65.

6. Mumey 1947, p. 310; Noble 1984, vol. 2, p. 163.

7. Walls 1989, p. 1; Ensminger 1992, pp. 178–80.

8. Clevinger 1938, 1942; Jordan 1985, pp. 32, 36–39.

9. Kilpinen 1994b; Knudsen 1969, pp. 7, 11, 33, 55, 62; Miller 1974, p. 66; Clemson 1978, p. 76; Clemson 1974, pp. 88, 93; Jordan 1993, pp. 293, 296.

10. Weis 1988, p. 261.

11. Jordan 1993, pp. 267–307.

12. Van Hoesen 1970, p. 5; one is preserved at the Matanuska Colony Village Museum (see appendix).

13. Sandoval 1986, pp. 42–43; Pitman 1973a, p. 224.

14. Sandoval 1986, pp. 41–42; Kilpinen 1995b.

15. Alspach 1994, pp. 65–71; Bealer and Ellis 1978, p. 20.

16. Jordan and Kaups 1989, pp. 219–21.

17. Moulton 1995.

18. Carlson 1990, p. 141; Gritzner 1990; Gritzner 1971, pp. 56–57; Stedman 1985, pp. 22, 30–31, 41, 44, 57–59; Jackson 1952, pp. 31–32; Wilson and Kammer 1989, pp. 33, 35, 90; Bunting 1962, p. 25.

19. Carlson 1990, pp. 139–41.

20. Carter 1978, p. 131; Russell 1984, p. 45; Sparling 1989, pp. 113, 122, 135; Kalman and de Visser 1976, pp. 160–61, 176–77; Veillete and White 1977, pp. 128, 170; Gulliford 1984, pp. 131, 166; Attebery 1991, p. 33; Bealer and Ellis 1978, pp. 174; Koskela 1985, p. 34; Read 1965, p. 163.

21. Walker 1984, pp. 139, 143, 146; Lavington 1982, pp. 2, 56.

22. An example is displayed at the Saamelais Museum, Inari, Finland.

23. Gritzner 1974b; Bunting 1976, p. 84. Several such Hispanic mills can be seen at Old Cienega Village Museum, Santa Fe, N.Mex. (see appendix).

Chapter Four Log Carpentry Traditions

1. Jordan 1985; Jordan and Kaups 1989, pp. 135–78.

2. Alspach 1994, pp. 155–58; Candee 1976.

3. Gritzner 1969, 1971, 1979–80; Winberry 1974; Bullock 1973; Bunting 1976; Jett and Spencer 1981.

4. Ahlborn 1967.

5. Koskela 1985; Carter 1984a; Knudsen 1969; Mattila 1971; Walls 1989; Lehr 1980; Wonders and Rasmussen 1980, pp. 209, 211, 215; Kilpinen 1994–95.

6. Jordan 1985, p. 14.

7. Jordan 1985, pp. 14–15.

8. Attebery 1976, p. 37; Roe 1958, p. 4; Jordan 1978, p. 36.

9. Clemson 1976, p. 96; Lacey 1990, p. 80; Clemson 1974, pp. 10, 26–27, 32, 44–46, 70, 79–81; Wonders and Rasmussen 1980, pp. 202–4.

10. Alspach 1994, pp. 22–28; Kilpinen 1994–95, pp. 29–30.

11. Jordan 1985, p. 46; Jordan 1978, p. 37.

12. Koskela 1985, p. 31; Jones 1978, p. 8; Carter 1984a; Wonders and

Rasmussen 1980, p. 215; Kilpinen 1994–95, p. 26; Sundre Pioneer Museum; Long Valley Museum (see appendix).

13. Roe 1958, p. 6; *Canadian Log House* 1974, p. 53.

14. Alspach 1994, pp. 29–34; *Canadian Log House* 1978, p. 78.

15. We use the terminology proposed in Kniffen 1969.

16. Jordan and Kilpinen 1990; Sultz 1964, p. 23; Jones 1978, p. 7; Walker 1984, p. 40; Wonders and Rasmussen 1980, pp. 203–4. We include the subtype known as "half-notching" in the square-notch type.

17. Davis 1989.

18. Jordan, Kaups, and Lieffort 1986; Jones 1978, p. 6; Sultz 1964, p. 23; Kildare 1967, p. 17; Wonders and Rasmussen 1980, pp. 203–4.

19. Jordan 1994a; Sheller 1957, precedes p. 33; Attebery 1976, p. 44; Walker 1984, p. 40; Weis 1988, p. 161; Roops Fort museum (see appendix).

20. Roe 1958, p. 4, described V-notching but called it "saddle." Winberry 1974, pp. 55–56, in an equally confusing way, described one subtype of saddle notching as "double"-notching. Walker 1984, pp. 40, 50, referred to the saddle type as "round" notching.

21. Jordan 1985, pp. 18–19; Lehr 1980, p. 187; Walker 1984, pp. xi, 40; Jones 1978, p. 6; Wonders and Rasmussen 1980, pp. 203–4.

22. Lavington 1982, follows p. 56.

23. Carter 1984a; Mattila 1971, p. 7; Kilpinen 1994–95, p. 26; Jones 1978, p. 8; Knudsen 1969, pp. 29–31; Sandoval 1986, p. 9; Koskela 1985, pp. 32–33; Clemson 1974, pp. 67–69; Brandt 1930, p. 180; Walker 1984, p. 41; Roe 1958, p. 3; Wonders and Rasmussen 1980, pp. 203–4, 211; Long Valley Museum; Pioneer Park, Ferndale (see appendix).

24. Alspach 1994, pp. 36, 43–49, 155–56.

25. Alspach 1994; pp. 36, 50–53.

26. Walker 1984, p. 124.

27. Gritzner 1971, pp. 58–59; Winberry 1974, pp. 55–56; Wonders and Rasmussen 1980, pp. 203–4. Winberry refers to the vertical single-notch as the "half-notch."

28. Carter 1984a; Koskela 1985, pp. 32–33; Kilpinen 1994–95, p. 26.

29. Jordan and Kilpinen 1990, pp. 14–17.

30. Bealer and Ellis 1978, pp. 162–63, 174; Reece 1985, pp. 20–23; Attebery 1991, pp. 16–17; Kalman and de Visser 1976, p. 178; Sultz 1964, pp. 26–27, 30; Kniffen and Glassie 1966, pp. 50–51; *Canadian Log House* 1975, pp. 48–49 and 1977, p. 47.

31. Kniffen and Glassie 1966, pp. 49–50; Stedman 1985, p. 39.

32. Kniffen and Glassie 1966, p. 52; Matanuska Colony Village Museum (see appendix).

33. Carter 1984a, pp. 64–65; Walker 1984, pp. 41, 124; Koskela 1985, pp. 32–33; Kilpinen 1994–95, p. 26.

34. Carlson 1977; Flaccus 1979; Mackie 1977, 1981; Schumann 1976; Walker 1984; *Canadian Log House* 1974–78.

35. Gritzner 1971, pp. 54–55; Wilson and Kammer 1989, p. 19.

36. Jones 1978, p. 4; Lehr 1980, pp. 186, 189.

37. Davidson 1977, p. 11; Jones 1978, p. 3; Sparling 1989, pp. 27, 46, 59, 84, 92–93; Read 1965, pp. 163, 165; Carter 1984a, pp. 63–64; Lavington 1982, follows p. 56.

38. Attebery 1976, p. 40.

39. Gritzner 1971, pp. 60–61; Attebery 1976, pp. 39, 45.

40. Clemson 1974, pp. 19, 32, 38, 42, 46–47, 81; Brown 1951, follows p. 80.

41. Everett 1966, p. 397.

42. Wonders and Rasmussen 1980, pp. 206–7.

43. Conway 1951, pp. 20–21; Wilson 1991; Gritzner 1971, pp. 60–62.

44. Stokes 1963, pp. 240–41; Wyckoff and Dilsaver 1995, p. 208.

45. Kilpinen 1994c; Kilpinen 1995b.

46. Baker 1983, p. 3; Paterson 1983, p. 7; Boag 1992, p. 57.

47. Carter 1978, p. 110; Pitman 1973a, p. 217; Weis 1988, pp. 129, 132; Wilson and Kammer 1989, p. 33; Attebery 1976, pp. 39, 42; Wonders and Rasmussen 1980, p. 206.

48. Drumheller 1925, p. 67; Everett 1966, pp. 50, 234; Attebery 1976, p. 39; Stedman 1985, p. 39; Roe 1958, p. 7; Wonders and Rasmussen 1980, p. 206.

49. Welsch 1980, pp. 324–25.

50. Clemson 1974, pp. 87–89.

Chapter Five Wooden Fences

1. Hart and Mather 1957, p. 4.

2. Mather and Hart 1954, p. 201.

3. The nucleus of this chapter is to be found in Kilpinen 1992. See also Clemson 1970 and Meredith 1951.

4. Clemson 1974, pp. 11–13, 52, 88; Clemson 1976, p. 103; Cline 1976, p. 183; Everett 1966, p. 40.

5. Fife 1988, p. 47.

6. Kennedy 1964, p. 95.

7. Jordan 1991, pp. 3, 6–7; Clemson 1974, pp. 58–59; Clemson 1976, pp. 13, 34; Kennedy 1964, p. 297.

8. Clemson 1974, pp. 49–56.

9. Jordan and Kaups 1989, pp. 109, 111, 112; Meredith 1951, pp. 131, 149, 150.

10. Clemson 1976, p. 106.

11. Rehder 1992, p. 117; Noble 1984, vol. 2, p. 122; Leechman 1953, pp. 226–27; Meredith 1951, p. 149.

12. Fife 1967, p. 51; Fife 1988, p. 47; Clemson 1974, p. 56; Clemson 1976, pp. 105–6; Downs 1975–79, vol. 2, p. 29.

13. Clemson 1970, p. 133; Clemson 1976, p. 106.

14. Clemson 1974, p. 57; Rikoon 1984, p. 62; Jordan and Kaups 1989, pp. 112–14.

15. Rikoon 1984, pp. 61–62; Jackson and Lee 1978, pp. 16–18; Jordan 1993, p. 288.

16. Sandoval 1986, pp. 14, 20; Burns et al. 1955, pp. 127, 134, 152, 154, 176, 199, 201; Kilpinen 1990, pp. 169–70.

17. Fife 1967, p. 52; Meredith 1951, p. 149.

18. Cassidy 1985, vol. 1, pp. 417–18; Hankey 1960, pp. 42–43.

19. Burns et al. 1955, p. 604.

20. Cassidy 1985, vol. 1, p. 406.

21. Meredith 1951, p. 125; Noble 1984, vol. 2, p. 130.

22. Brown 1975, p. 112.

23. Fife 1967, follows p. 52.

24. Clemson 1974, p. 55; Clemson 1976, p. 74; Burns et al. 1955, pp. 192, 201.

25. Hewes 1982, p. 326.

26. Lavington 1982, p. 216; Noble 1984, vol. 2, pp. 129–30; Leechman 1953, p. 227.

27. Clemson 1974, p. 55.

28. Clemson 1970, p. 132.

29. Clemson 1974, pp. 14, 52, 55, 57; Clemson 1970, pp. 132–33; see also Downs 1975-79, pp. 27–28; Leechman 1953, p. 227.

30. Clemson 1974, p. 55.

31. Clemson 1970, pp. 131–32.

32. Clemson 1970, p. 132; Clemson 1974, p. 55; Downs 1975–79, p. 27.

33. Vaughan and Ferriday 1974, pp. 253–54; Rikoon 1984, pp. 62–64.

34. Francaviglia 1978b, pp. 29, 76; Noble 1984, vol. 2, pp. 131–32.

Chapter Six Material Culture of Haymaking

1. Jordan and Kaups 1989.

2. Jordan 1993, p. 302.

3. McCormick et al. 1979, p. 203.

4. Hurt 1982, p. 92.

5. Young 1983.

6. Kilpinen 1991, p. 34.

7. Young 1983, p. 311.

8. Young 1983, p. 318.

9. Armitage 1955, pp. 204–6.

10. Young 1983, pp. 319–20.

11. Oliver and Jackman 1962, p. 160.

12. Alwin 1982; Noble 1984, vol. 2, pp. 112–14.

13. Armitage 1955, pp. 206–7.

14. Alwin 1982, p. 47.

15. Burns et al. 1955, pp. 241, 605; Sandoval 1986, p. 20; Crowley 1964, pp. 143–47; Armitage 1955, p. 207.

16. Armitage 1955, p. 208; Alwin 1982, p. 42.

17. Alwin 1982, pp. 44, 48–49.

18. Alwin 1982, p. 47.

19. Armitage 1955, p. 210.

20. Armitage 1955, p. 207.

21. Francaviglia 1978a, p. 922; Noble 1984, vol. 2, pp. 110–13, 132; Lavington 1982, follows p. 86.

22. Fife and Fife 1948, 1951; Francaviglia 1978a, pp. 922–23.

23. Francaviglia 1978a, p. 919.

24. Fife and Fife 1948, 1951; Francaviglia 1978a.

25. Francaviglia 1978a, pp. 923–25.

26. Francaviglia 1978a, p. 923.

27. Young 1983, p. 313.

28. Alwin 1982, p. 46.

29. Young 1983, pp. 316–17.

30. Young 1983, pp. 313, 320.

31. Hurt 1982, pp. 95–96; Everett 1966, p. 429; Alwin 1982, p. 45; Burns et al. 1955, p. 161.

32. Fife and Fife 1948, 1951.
33. Young 1983.
34. Mather and Hart 1954, p. 218.
35. Nowlin 1883, p. 504; Jordan and Kaups 1989, p. 99.
36. Wacker 1973, pp. 38–39, 41–44.
37. Young and Sparks 1985, p. 194.
38. Hankey 1961, p. 268.

Chapter Seven **The West Revealed**

1. Jordan 1989
2. Jordan 1993, pp. 268, 299–306.
3. Lee and Lee 1981.
4. Meinig 1972; Meinig 1965; Carlson 1990; Francaviglia 1978b.

References

Ahlborn, Richard E. 1967. "The Wooden Walls of Territorial New Mexico." *New Mexico Architecture* 9 (9–10): 20–23.

Alspach, Elizabeth K. 1994. "Reading Ontario's Folk Landscape: The Province's American Roots Revealed in Pioneer Log Structures." M.A. thesis. University of Texas, Austin.

Alwin, John A. 1982. "Montana's Beaverslide Hay Stacker." *Journal of Cultural Geography* 3 (1): 42–50.

Armitage, Herbert S. 1955. "Haying Operations in the Big Hole." In *The Land of the Big Snows*, edited by Bertha A. Francis, pp. 203–12. Caldwell, Idaho: Caxton Printers.

Arthur, Eric, and Witney, Dudley. 1972. *The Barn: A Vanishing Landmark in North America*. Toronto: M. F. Feheley.

Attebery, Jennifer E. 1976. "Log Construction in the Sawtooth Valley of Idaho." *Pioneer America* 8 (1): 36–46.

———. 1982. "The Square Cabin: A Folk House Type in Idaho." *Idaho Yesterdays* 26 (Fall): 25–31.

———. 1985. "The Diffusion of Folk Culture as Demonstrated in the Horizontal Timber Construction of the Snake River Basin." Ph.D. dissertation. Indiana University, Bloomington.

———. 1991. *Building Idaho: An Architectural History*. Moscow: University of Idaho Press.

Attebery, Louie W., ed. 1985. *Idaho Folklife: Homesteads to Headstones*. Salt Lake City: University of Utah Press, and Boise: Idaho State Historical Society.

August, Ray. 1993. "Cowboys v. Rancheros: The Origins of Western American Livestock Law." *Southwestern Historical Quarterly* 96: 457–88.

Baker, Byrd. 1983. *Mendocino Past and Present*. Mendocino, Calif.: Pacific Transcriptions.

Bealer, Alex W., and Ellis, John O. 1978. *The Log Cabin: Homes of the North American Wilderness.* Barre, Mass.: Barre Publishing.

Bloom, J. T. 1972. "Cumberland Gap versus South Pass: The East or West in Frontier History." *Western Historical Quarterly* 3: 153–67.

Boag, Peter G. 1992. *Environment and Experience: Settlement Culture in Nineteenth–Century Oregon.* Berkeley and Los Angeles: University of California Press.

Boyd, E. 1958. "Fireplaces and Stoves in Colonial New Mexico." *El Palacio* 65 (6): 219–24.

Brandt, Lucas. 1930. "Pioneer Days on the Big Thompson." *Colorado Magazine* 7: 178–82.

Brown, Kimberly R. 1975. *Historical Overview of the Dillon District.* Boulder: Western Interstate Commission for Higher Education.

Brown, Robert L. 1990. *Jeep Trails to Colorado Ghost Towns.* Caldwell, Idaho: Caxton.

Brown, William S. 1951. *California Northeast: The Bloody Ground.* Oakland, Calif.: Biobooks.

Brumbaugh, G. Edwin. 1933. "Colonial Architecture of the Pennsylvania Germans." *Pennsylvania German Society Proceedings* 41(2): 5–60.

Bullock, Alice. 1973. *Mountain Villages.* Santa Fe: Sunstone Press.

Bunting, Bainbridge. 1962. "The Architecture of the Embudo Watershed." *New Mexico Architecture* 4: 25.

———. 1976. *Early Architecture in New Mexico.* Albuquerque: University of New Mexico Press.

Bunting, Bainbridge, and Conron, J. P., 1966. "The Architecture of Northern New Mexico." *New Mexico Architecture* 8 (Sep.–Oct.): 14–49.

Burns, Robert H.; Gillespie, Andrew S.; and Richardson, Willing G. 1955. *Wyoming's Pioneer Ranches.* Laramie, Wyo.: Top-of-the-World Press.

The Canadian Log House. 1974–78. vols. 1–5 (a yearbook edited by B. Allan Mackie) Prince George, British Columbia.

Candee, Richard M. 1976. "Wooden Buildings in Early Maine and New Hampshire." Ph.D. dissertation. University of Pennsylvania, Philadelphia.

Carlson, Alvar W. 1990. *The Spanish-American Homeland: Four Centuries in New Mexico's Río Arriba.* Baltimore: Johns Hopkins University Press.

Carlson, Axel R. 1977. *Building a Log House in Alaska.* Fairbanks: University of Alaska Cooperative Extension Service.

Carter, Thomas. 1984a. "North European Horizontal Log Construction in the Sanpete-Sevier Valleys." *Utah Historical Quarterly* 52: 50–71.

———. 1984b. "Building Zion: Folk Architecture in the Mormon Settlements of Utah's Sanpete Valley." Ph.D. dissertation. Indiana University, Bloomington.

Carter, William. 1978. *Ghost Towns of the West.* Menlo Park, Calif.: Lane Publishing.

Cassidy, Frederic G., ed. 1985. *Dictionary of American Regional English.* Cambridge: Harvard University Press.

Chappel, Jill A. 1990. "Homestead Ranches of the Fort Rock Valley: Vernacular Building in the Oregon High Desert." M.S. thesis. University of Oregon, Eugene.

Clemson, Donovan. 1970. "Some Western Fences." *Canadian Geographical Journal* 81: 130–35.

———. 1974. *Living with Logs: British Columbia's Log Buildings and Rail Fences.* Saanichton, British Columbia: Hancock House.

———. 1976. *Outback Adventures: Through Interior British Columbia.* Saanichton, British Columbia: Hancock House.

———. 1978. *Old Wooden Buildings.* Seattle: Hancock House.

Clevinger, W. R. 1938. "The Appalachian Mountaineers in the Upper Cowlitz Basin." *Pacific Northwest Quarterly* 29: 115–34.

———. 1942. "Southern Appalachian Highlanders in Western Washington." *Pacific Northwest Quarterly* 33: 3–25.

Cline, Platt. 1976. *They Came to the Mountain.* Flagstaff: Northern Arizona University and Northland Press.

Conway, A. W. 1951. "A Northern New Mexico House-Type." *Landscape* 1 (Autumn): 20–21.

Crowley, John M. 1964. "Ranches in the Sky: A Geography of Livestock Ranching in the Mountain Parks of Colorado." Ph.D. dissertation. University of Minnesota, Minneapolis.

Dallas, Sandra. 1981. *Colorado Homes.* Norman: University of Oklahoma Press.

———. 1985. *Colorado Ghost Towns and Mining Camps.* Norman: University of Oklahoma Press.

Davidson, David. 1977. "Log Buildings in the San Francisco Peaks Area of Northern Arizona." *Southwest Folklore* 1: 1–28.

Davis, Kathleen A. 1989. "A New England Log House on the Central California Coast." *Material Culture* 21 (3): 27–39.

Downs, Art, ed. 1975-79. *Pioneer Days in British Columbia.* Surrey, British Columbia: Heritage House.

Drumheller, Daniel M. 1925. *Uncle Dan Drumheller Tells Thrills of Western Trails.* Spokane: Inland-American Printing.

Engbeck, Joseph H., Jr., ed. 1990. *California Historical Landmarks.* Sacramento: California Dept. of Parks and Recreation, Office of Historic Preservation.

Ensminger, Robert F. 1992. *The Pennsylvania Barn: Its Origin, Evolution, and Distribution in North America.* Baltimore: Johns Hopkins University Press.

Evans, Timothy. 1988. *Wyoming's Material Folk Arts.* Laramie: University of Wyoming, American Studies Program.

Everett, George C. 1966. *Cattle Cavalcade in Central Colorado.* Denver: Golden Bell Press.

Fife, Alta, ed. 1988. *Exploring Western Americana.* Ann Arbor: UMI Research Press.

Fife, Austin E. 1957. "Folklore of Material Culture on the Rocky Mountain Frontier." *Arizona Quarterly* 13: 101–10.

———. 1967. "Jack Fences of the Intermountain West." In *Folklore In-*

ternational: Essays in Traditional Literature, Belief, and Custom in Honor of Wayland Debs Hand,* edited by D. K. Wilgus, pp. 51–54. Hatboro, Pa.: Folklore Associates.

Fife, Austin E., and Fife, James M. 1948, 1951. "Hay Derricks of the Great Basin and Upper Snake River Valley." *Western Folklore* 7: 225–39; 10: 320–22.

Flaccus, Edward. 1979. *North Country Cabin.* Missoula, Mont.: Mountain Press.

Francaviglia, Richard V. 1972. "Western American Barns: Architectural Form and Climatic Considerations." *Yearbook of the Association of Pacific Coast Geographers* 34: 153–60.

———. 1978a. "Western Hay Derricks: Cultural Geography and Folklore as Revealed by Vanishing Agricultural Technology." *Journal of Popular Culture* 11 (Spring): 916–27.

———. 1978b. *The Mormon Landscape: Existence, Creation, and Perception of a Unique Image in the American West.* New York: AMS Press.

Freed, Elaine. 1992. *Preserving the Great Plains and Rocky Mountains.* Albuquerque: University of New Mexico Press.

Freeman, Ira S. 1958. *A History of Montezuma County, Colorado.* Boulder: Johnson.

Glover, Margaret L. 1982a. "Log Structures: Criteria for Their Description, Evaluation and Management as Cultural Resources." M.A. thesis. Portland State University, Portland, Oreg.

———. 1982b. "Horizontal Log Construction Corner Types." *Northwest Anthropological Research Notes* 16 (2): 165–85.

Gray, Edward. 1989. *An Illustrated History of Early Northern Klamath County, Oregon.* Bend, Oreg.: Maverick.

Gressley, Gene M. 1986. "The West: Past, Present, and Future." *Western Historical Quarterly* 17: 4–23.

Gritzner, Charles F. 1969. "Spanish Log Construction in New Mexico." Ph.D. dissertation. Louisiana State University, Baton Rouge.

———. 1971. "Log Housing in New Mexico." *Pioneer America* 3 (2): 54–62.

———. 1974a. "Construction Materials in a Folk Housing Tradition: Considerations Governing Their Selection in New Mexico." *Pioneer America* 6 (1): 25–39.

———. 1974b. "Hispano Gristmills in New Mexico." *Annals of the Association of American Geographers* 64: 514–24.

———. 1979-80. "Hispanic Log Construction of New Mexico." *El Palacio* 85: 20–29.

———. 1990. "Log Barns of Hispanic New Mexico." *Journal of Cultural Geography* 10 (2): 21–34.

Gulliford, Andrew. 1984. *America's Country Schools.* Washington, D.C.: Preservation Press.

Hankey, Clyde. 1960. "A Colorado Word Geography." *Publications of the American Dialect Society* 34: entire issue.

———. 1961. "Semantic Features of Eastern Relics in Colorado Dialect." *American Speech* 36: 266–70.

Hart, John F., and Mather, E. Cotton. 1957. "The American Fence." *Landscape* 6 (3): 4–9.

Hartung, John. 1978. "Documentation of the Historical Resources in the Idaho Primitive Area, Big Creek Drainage." M.A. thesis. University of Idaho, Moscow.

Hewes, Leslie. 1982. "Early Fencing on the Western Margin of the Prairie." *Nebraska History* 63: 301–48.

Hoesen, Diana van. 1970. "First Barns of the Matanuska Valley." *This Alaska* 2 (5): 5–7, 28.

Hudson, John C. 1975. "Frontier Housing in North Dakota." *North Dakota History* 42 (4): 4–15.

Hurt, R. Douglas. 1982. "American Farm Tools: From Hand-Power to Steam-Power." *Journal of the West* 21 (1): 1–112.

Jackson, John B. 1952. "A Catalog of New Mexico Farm-Building Terms." *Landscape* 1 (3): 31–32.

———. 1994. *A Sense of Place, a Sense of Time.* New Haven: Yale University Press.

Jackson, Royal, and Lee, Jennifer. 1978. *Harney County: An Historical Inventory.* Burns, Oreg.: Harney County Historical Society.

Jenkinson, Michael, and Kernberger, Karl. 1968. *Ghost Towns of New Mexico.* Albuquerque: University of New Mexico Press.

Jett, Stephen C., and Spencer, Virginia E. 1981. *Navajo Architecture: Forms, History, Distributions.* Tucson: University of Arizona Press.

Jones, Larry. 1978. "Utah's Vanishing Log Cabins." *Beehive History* 4: 3–9.

Jordan, Teresa. 1992. *Cowgirls: Women of the American West.* Lincoln: University of Nebraska Press.

Jordan, Terry G. 1978. *Texas Log Buildings: A Folk Architecture.* Austin: University of Texas Press.

———. 1985. *American Log Buildings.* Chapel Hill: University of North Carolina Press.

———. 1989. "Preadaptation and European Colonization in Rural North America." *Annals of the Association of American Geographers* 79: 489–500.

———. 1991. "The North American West: Continuity or Innovation?" In *The Dauphin Papers: Research by Prairie Geographers,* edited by John Welsted and John Everitt, pp. 1–17. Brandon, Manitoba: Brandon University Geographical Studies no. 1.

———. 1993. *North American Cattle-Ranching Frontiers.* Albuquerque: University of New Mexico Press.

———. 1994a. "Crowned V-Notching in the Midland American Log Culture Complex." *Pioneer America Society Transactions* 17: 19–23.

———. 1994b. "The 'Saddlebag' House Type and Pennsylvania Extended." *Pennsylvania Folklife* 44 (1): 36–48.

Jordan, Terry G.; Domosh, Mona; and Rowntree, Lester. 1994. *The Human Mosaic.* 6th ed. New York: HarperCollins.

Jordan, Terry G., and Kaups, Matti. 1987. "Folk Architecture in Cultural and Ecological Context." *Geographical Review* 77: 52–75.

———. 1989. *The American Backwoods Frontier.* Baltimore: Johns Hopkins University Press.

Jordan, Terry G.; Kaups, Matti; and Lieffort, Richard M. 1986. "New Evidence on the European Origin of Pennsylvanian V Notching." *Pennsylvania Folklife* 36 (1): 20–31.

Jordan, Terry G., and Kilpinen, Jon T. 1990. "Square Notching in the Log Carpentry Tradition of Pennsylvania Extended." *Pennsylvania Folklife* 40 (1): 2–18.

Kalman, Harold, and de Visser, John. 1976. *Pioneer Churches.* New York: W. W. Norton.

Kariel, Herbert G. 1992. "Alpine Huts in Canada's Western Mountains." *Canadian Geographer* 36: 144–58.

Kennedy, Michael S. 1964. *Cowboys and Cattlemen.* New York: Hastings House.

Kersten, Earl W., Jr. 1964. "The Early Settlement of Aurora, Nevada, and Nearby Mining Camps." *Annals of the Association of American Geographers* 54: 490–507.

Kildare, Maurice. 1967. "Old Cabin Corners." *Relics* 1 (2): 16–18.

———. 1970. "Cabins West." *Relics* 3 (5): 6–9, 28–29.

Kilpinen, Jon T. 1990. "Material Folk Culture in the Adaptive Strategy of the Rocky Mountain Valley Ranching Frontier." M.A. thesis. University of Texas, Austin.

———. 1991. "Material Folk Culture of the Rocky Mountain High Valleys." *Material Culture* 23 (2): 25–41.

———. 1992. "Traditional Fence Types of Western North America." *Pioneer America Society Transactions* 15: 15–22.

———. 1994a. "The Origins of Traditional Mountain Western Culture as Revealed in the Built Landscape." Ph.D. dissertation. University of Texas, Austin.

———. 1994b. "The Mountain Horse Barn: A Case of Western Innovation." *Pioneer America Society Transactions* 17: 25–32.

———. 1994c. "Cultural Diffusion and the Formation of the Western Cultural Landscape: The Shed-Roofed Single-Crib Outbuilding." *Program and Paper Abstracts, 1994 Pioneer America Society Conference,* November 3–6, 1994, p. 23.

———. 1994–95. "Finnish Cultural Landscapes in the Pacific Northwest." *Pacific Northwest Quarterly* 86 (1): 25–34.

———. 1995a. "The Front-Gabled Log Cabin and the Role of the Great Plains in the Formation of the Mountain West's Built Landscape." *Great Plains Quarterly* 15 (1): 19–31.

———. 1995b. "Cultural Diffusion and the Formation of the Western Cultural Landscape: The Shed-Roofed Single-Crib Outbuilding." *Pioneer America Society Transactions* 18: 9–15.

Kniffen, Fred B. 1960. "To Know the Land and Its People." *Landscape* 9: 20–23.

———. 1965. "Folk Housing: Key to Diffusion." *Annals of the Association of American Geographers* 55: 549–77.

———. 1969. "On Corner Timbering." *Pioneer America* 1 (1): 1–8.

Kniffen, Fred B., and Glassie, Henry. 1966. "Building in Wood in the Eastern United States." *Geographical Review* 56: 40–66.

Knudsen, Harold S. 1969. "Barns as an Index to Ethnic Origins in Western Montana." M.A. thesis. University of Montana, Missoula.

Koskela, Alice. 1985. "Finnish Log Homestead Buildings in Long Valley." In *Idaho Folklife: Homesteads to Headstones,* edited by Louie W. Attebery, pp. 29–36. Salt Lake City: University of Utah Press, and Boise: Idaho State Historical Society.

Krenzlin, Anneliese. 1965. *Die Agrarlandschaft an der Nordgrenze der Besiedlung im intermontanen British Columbia.* Frankfurt-am-Main: Waldemar Kramer.

Lacey, Donna M. 1990. *Photographing Montana, 1894–1928.* New York: Alfred A. Knopf.

Lavington, H. 1982. *The Nine Lives of a Cowboy.* Victoria, British Columbia: Sono Nis Press.

Lee, David R., and Lee, Hector H. 1981. "Thatched Cowsheds of the Mormon Country." *Western Folklore* 40: 171–87.

Leechman, Douglas. 1953. "Good Fences Make Good Neighbors." *Canadian Geographical Journal* 47: 218–35.

Lehr, John C. 1980. "The Log Buildings of Ukrainian Settlers in Western Canada." *Prairie Forum* 2: 183–96.

Limerick, Patricia N. 1987. *The Legacy of Conquest: The Unbroken Past of the American West.* New York: W. W. Norton.

Limerick, Patricia N.; Milner, Clyde A.; and Rankin, Charles, eds. 1991. *Trails: Toward the "New" Western History.* Lawrence: University Press of Kansas.

Lufkin, Agnesa Burney. 1985. "Domestic Architecture in Northeastern New Mexico, Late Territorial Period, 1880–1912." Ph.D. dissertation. University of New Mexico, Albuquerque.

McCormick, John; Young, James A.; and Burkhardt, Wayne. 1979. "Making Hay." *Rangelands* 1 (5): 203–6.

McDaniel, Marion, and Wylie, Jerry. 1979. *Cultural Resource Evaluation of Sater Cabin—Little Creek (Middle Fork Ranger District, Challis National Forest, Idaho).* Ogden, Utah: U.S. Dept. of Agriculture, Forest Service, Intermountain Region.

Mackie, B. Allan. 1977. *Notches of All Kinds: A Book of Timber Joinery.* Prince George, British Columbia: Hairy Woodpecker.

————. 1981. *Building with Logs.* New York: Charles Scribner's Sons.

Mann, Dale, and Skinulis, Richard. 1979. *The Complete Log House Book: A Canadian Guide to Building with Logs.* San Francisco: McGraw-Hill Ryerson.

Mather, E. Cotton, and Hart, John F. 1954. "Fences and Farms." *Geographical Review* 44: 201–23.

Matheson, Janet. 1985. *Fairbanks: A City Historic Buildings Survey.* Fairbanks: City of Fairbanks.

Mattila, Walter. 1971. "The Pioneer Finnish Home." *Finnish-American Historical Society of the West* 6 (2): 1–23.

Meinig, Donald W. 1965. "The Mormon Culture Region." *Annals of the Association of American Geographers* 55: 191–220.

————. 1972. "American Wests: Preface to a Geographical Interpretation." *Annals of the Association of American Geographers* 62: 159–84.

————, ed. 1979. *The Interpretation of Ordinary Landscapes.* New York: Oxford University Press.

Meitzen, August. 1882. *Das deutsche Haus.* Berlin: Dietrich Reimer.

————. 1895. *Siedelung und Agrarwesen der Westgermanen und Ostgermanen, der Kelten, Römer, Finnen und Slawen.* Berlin: Wilhelm Hertz.

Meredith, Mamie. 1951. "The Nomenclature of American Pioneer Fences." *Southern Folklore Quarterly* 15: 109–51.

Miller, Donald C. 1974. *Ghost Towns of Montana.* Boulder: Pruett Publishing.

Monaghan, Robert L. 1959. "The Development of Settlement in the Fairbanks Area: A Study of Permanence." Ph.D. dissertation. McGill University, Montreal.

Moulton, Candy. 1995. "Jackson Hole Barn Stands Strong Once Again." *Persimmon Hill* 23 (3): 10–11.

Mumey, Nolie. 1947. *The Teton Mountains: Their History and Tradition.* Denver: Artcraft Press.

Nelson, Richard K. 1986. *Hunters of the Northern Forest: Designs for Survival among the Alaskan Kutchin.* 2d ed. Chicago: University of Chicago Press.

Noble, Allen G. 1984. *Wood, Brick, and Stone: The North American Settlement Landscape.* 2 vols. Amherst: University of Massachusetts Press.

Nowlin, William. 1883. "The Bark-Covered House, or Pioneer Life in Michigan." *Report of the Pioneer Society of Michigan* 4: 480–541.

Oliver, Herman, and Jackman, E. R. 1962. *Gold and Cattle Country.* 2d ed. Portland, Oreg.: Binfords & Mort.

Paher, Stanley W. 1970. *Nevada Ghost Towns and Mining Camps.* Berkeley, Calif.: Howell-North.

Parker, Watson, and Lambert, Hugh K. 1974. *Black Hills Ghost Towns.* Chicago: Swallow Press.

Paterson, T. W. 1983. *British Columbia Ghost Town Series: Vancouver Island.* Langley, British Columbia: Sunfire.

Perry, Edgar. 1971. *The Old Log Cabin.* Tucson: Arizona Historical Society.

Pitman, Leon S. 1973a. "A Survey of Nineteenth-Century Folk Housing in the Mormon Culture Region." Ph.D. dissertation. Louisiana State University, Baton Rouge.

————. 1973b. *A Survey of the Nineteenth-Century Folk Housing of the Mormons.* Salt Lake City: Utah State Historical Society.

Pomeroy, Earl. 1955. "Toward a Reorientation of Western History: Continuity and Environment." *Mississippi Valley Historical Review* 41: 579–600.

Poulsen, Richard C. 1982. *The Pure Experience of Order: Essays on the Symbolic in the Folk Material Culture of Western America.* Albuquerque: University of New Mexico Press.

Ramsey, Bruce. 1963. *Ghost Towns of British Columbia.* Vancouver, British Columbia: Mitchell Press.

Read, Effie O. 1965. *White Pine Lang Syne: A True History of White Pine County, Nevada*. Denver: n.p.

Reece, Daphne. 1985. *Historic Houses of the Pacific Northwest*. San Francisco: Chronicle Books.

Reed, Adele, and Smith, Genny. 1982. *Old Mammoth*. Mammoth Lakes, Calif.: Genny Smith.

Reeve, Agnes Lufkin. 1988. *From Hacienda to Bungalow: Northern New Mexico Houses, 1850–1912*. Albuquerque: University of New Mexico Press.

Rehder, John B. 1992. "The Scotch-Irish and English in Appalachia." In *To Build in a New Land*, edited by Allen G. Noble, pp. 95–118. Baltimore: Johns Hopkins University Press.

Reitzes, Lisa B. 1981. *Paris: A Look at Idaho Architecture*. Boise: Idaho State Historic Preservation Office.

Rempel, John I. 1980. *Building with Wood and Other Aspects of Nineteenth-Century Building in Central Canada*. Rev. ed. Toronto: University of Toronto Press.

Reynolds, Brad, and Doll, Don. 1990. "Athapascans along the Yukon." *National Geographic* 177 (2): 44–69.

Rikoon, J. Sanford. 1984. "Traditional Fence Patterns in Owyhee County, Idaho." *Pioneer America Society Transactions* 7: 59–69.

Robbins, William G. 1986. "The 'Plundered Province' Thesis and the Recent Historiography of the American West." *Pacific Historical Review* 55: 577–97.

Rock, James T. 1980. *Horizontal Log Construction*. Happy Camp, Calif.: Siskiyou County Historical Society, Technical Series no. 1.

Roe, Frank G. 1958. "The Old Log House in Western Alberta." *Alberta Historical Review* 6 (2): 1–9.

Russell, Bert. 1979. *Swiftwater People*. Harrison, Idaho: Lacon.

———. 1984. *North Fork of the Coeur d' Alene River*. Harrison, Idaho: Lacon.

Sandoval, Judith H. 1986. *Historic Ranches of Wyoming*. Casper, Wyo.: Mountain States Lithographing and Nicolaysen Art Museum.

Sauer, Carl O. 1930. "Historical Geography and the Western Frontier." In *The Trans-Mississippi West*, edited by James F. Willard and Colin B. Goodykoontz, pp. 267–89. Boulder: University of Colorado Press.

Schar, Kenneth, and Pearl, George C. 1973. "Historic Structures of Catron County." *El Palacio* 79 (2): 3–15, cover illus.

Schumann, David R. 1976. *Building with House Logs in Alaska*. N.p.: U.S. Dept. of Agriculture, Forest Service, Alaska Region.

Segger, Martin. 1977. *Log Idiom Survivals in West Coast Architecture*. Banff, Alberta: Canadian Log Structures Conference.

Sheller, Roscoe. 1957. *Ben Snipes, Northwest Cattle King*. Portland, Oreg.: Binfords & Mort.

Simpich, Frederick. 1923. "Missouri, Mother of the West." *National Geographic* 43: 421–60.

Smith, Duane A. 1967. *Rocky Mountain Mining Camps*. Bloomington: Indiana University Press.

Sparling, Wayne. 1989. *Southern Idaho Ghost Towns.* Caldwell, Idaho: Caxton.

Stedman, Myrtle. 1985. *Rural Architecture of Northern New Mexico and Southern Colorado.* Santa Fe: Westamerica Publishing.

Stokes, George A. 1963. "The Pyramidal House." *Louisiana Studies* 2: 240–41.

Sultz, Philip W. 1964. "From Sagebrush to Hay and Back Again." *American West* 1 (1): 20–30.

———. 1969. "Architectural Values of Early Frontier Log Structures." In *Forms upon the Frontier: Folklife and Folk Arts in the United States,* edited by Austin E. Fife, Alta Fife, and Henry H. Glassie, pp. 31–40. Logan: Utah State University Press.

Turner, Frederick J. 1893. "The Significance of the Frontier in American History." *Annual Report of the American Historical Association,* pp. 199–227.

Turner, Frederick. 1990. *Of Chiles, Cacti, and Fighting Cocks: Notes on the American West.* San Francisco: North Point Press.

United States Dept. of Agriculture. 1939. *The Farm-Housing Survey.* Miscellaneous Publication no. 323. Washington, D.C.: Government Printing Office.

Vaughan, Thomas, and Ferriday, Virginia G., eds. 1974. *Space, Style and Structure: Buildings in Northwest America.* 2 volumes. Portland: Oregon Historical Society.

Veillette, John, and White, Gary. 1977. *Early Indian Village Churches: Wooden Frontier Architecture in British Columbia.* Vancouver: University of British Columbia Press.

Wacker, Peter O. 1973. "Folk Architecture as an Indication of Culture Areas and Culture Diffusion: Dutch Barns and Barracks in New Jersey." *Pioneer America* 5 (2): 37–47.

Walker, Tom. 1984. *Building the Alaska Log Home.* Anchorage: Alaska Northwest Publishing.

Walls, Robert E. 1989. "Ethnicity and Architecture in Eastern Washington: A Legacy of Continuity and Change." *Wash Board: The Newsletter of the Washington State Folklife Council* 5 (1): 1.

Webb, Walter P. 1931. *The Great Plains.* Boston: Ginn & Co.

Weis, Norman D. 1988. *Ghost Towns of the Northwest.* Caldwell, Idaho: Caxton.

Welsch, Roger L. 1980. "Nebraska Log Construction: Momentum in Tradition." *Nebraska History* 61: 310–35.

White, Richard. 1991. *"It's Your Misfortune and None of My Own": A History of the American West.* Norman: University of Oklahoma Press.

Wilson, Chris. 1991. "Pitched Roofs over Flat: The Emergence of a New Building Tradition in Hispanic New Mexico." In *Perspectives in Vernacular Architecture,* IV, edited by Thomas Carter and Bernard L. Herman. Columbia: University of Missouri Press.

Wilson, Chris, and Kammer, David. 1989. *Community and Continuity: The History, Architecture and Cultural Landscape of La Tierra Amarilla.* Santa Fe: New Mexico Historic Preservation Division.

Wilson, Mary. 1984. *The Rocky Mountain Cabin* (Part 1 of *Log Cabin*

Studies). Ogden, Utah: U.S. Dept. of Agriculture, Forest Service, Intermountain Division.

Winberry, John J. 1974. "The Log House in Mexico." *Annals of the Association of American Geographers* 64: 54–69.

Wonders, William C. 1979. "Log Dwellings in Canadian Folk Architecture." *Annals of the Association of American Geographers* 69: 187–207.

Wonders, William C., and Rasmussen, Mark A. 1980. "Log Buildings of West Central Alberta." *Prairie Forum* 5 (2): 197–217.

Woodward, Claire V. 1975. "Ethnohistory of Baker Cabin, Clackamas City, Oregon." M.A. thesis. Portland State University, Portland, Oreg.

Worster, Donald. 1985. *Rivers of Empire: Water, Aridity, and the Growth of the American West.* New York: Oxford University Press.

———. 1987. "New West, True West: Interpreting the Region's History." *Western Historical Quarterly* 18: 141–56.

———. 1992. *Under Western Skies: Nature and History in the American West.* New York: Oxford University Press.

Wunder, John R. 1994. "What's Old about the New Western History: Race and Gender." *Pacific Northwest Quarterly* 85: 50–58.

Wyckoff, William, and Dilsaver, Lary M., eds. 1995. *The Mountainous West: Explorations in Historical Geography.* Lincoln: University of Nebraska Press.

Wylie, Jerry. 1978. *Cultural Resource Evaluation of the Sack (Kipp) Cabin, Big Springs.* (Island Park Ranger Dist., Targhee Nat. Forest, Idaho). Ogden, Utah: U.S. Dept. of Agriculture, Forest Service, Intermountain Region.

———. 1979. *A Cultural Resource Evaluation of the Cabin Creek Ranch, Payette National Forest.* Ogden, Utah: U.S. Dept. of Agriculture, Forest Service, Intermountain Region.

Young, James A. 1983. "Hay Making: The Mechanical Revolution on the Western Range." *Western Historical Quarterly* 14: 311–26.

Young, James, and Sparks, B. Abbott. 1985. *Cattle in the Cold Desert.* Logan: Utah State University Press.

Young, Mary. 1970. "The West and American Cultural Identity: Old Themes and New Variations." *Western Historical Quarterly* 1: 137–60.

Yukon Historical and Museums Association. 1983. *Whitehorse Heritage Buildings.* Vancouver, British Columbia: Evergreen Press.

Index

Terry G. Jordan is the Walter Prescott Webb Professor of History and Ideas in the Department of Geography at the University of Texas in Austin. He is the author or coauthor of more than a dozen books, among them *American Log Buildings: An Old World Heritage* (North Carolina, 1985), *The American Backwoods Frontier: An Ethnic and Ecological Interpretation* (Johns Hopkins, 1989), and *North American Cattle Ranching Frontiers: Origins, Diffusion, and Differentiation* (New Mexico, 1993). Among his many decorations is the first John Wesley Powell Medal, a lifetime achievement award given by the New Mexico Geographical Society to that scholar, writer, or artist whose contributions to geographical knowledge are exemplary.

Jon T. Kilpinen is a faculty member in the Department of Geography at Valparaiso University. He has published widely on the material culture and cultural geography of the mountain West in professional journals.

Charles F. Gritzner is the Distinguished Service Professor of Geography at South Dakota State University. He has published seminal articles on the material culture and Hispanos of Highland New Mexico in professional journals, and is the author or coauthor of five books, among them *World Geography* (Heath, 1987, 1989, 1991) and *Exploring Our World: Past and Present* (Heath, 1991).

Library of Congress Cataloging-in-Publication Data

Jordan, Terry G.
    The Mountain West : interpreting the folk landscape /
Terry G. Jordan, Jon T. Kilpinen, Charles F. Gritzner.
        p.   cm. — (Creating the North American landscape)
    Includes bibliographical references and index.
    ISBN 0-8018-5431-8 (acid-free paper)
    1. Material culture—West (U.S.)   2. Log buildings—West
(U.S.)—Design and construction.   3. Vernacular architec-
ture —West (U.S.)   4. Landscape assessment—West (U.S.)
5. West (U.S.)—Geography.   I. Kilpinen, Jon T.   II. Gritzner,
Charles F.   III. Title.   IV. Series.
F590.7.J67   1997
917.8—dc20   96-15912
                                                            CIP